# THE MEANING OF ILLNESS

# Philosophy and Medicine

## VOLUME 42

### *Editors*

H. Tristram Engelhardt, Jr., *Center for Ethics, Medicine, and Public Issues, Baylor College of Medicine, Houston, Texas and Philosophy Department, Rice University, Houston, Texas*

Stuart F. Spicker, *School of Medicine, University of Connecticut Health Center, Farmington, Connecticut*

### *Editorial Board*

# THE MEANING
# OF ILLNESS

*A Phenomenological Account of the
Different Perspectives of Physician and Patient*

*by*

**S. KAY TOOMBS**

*Department of Philosophy, Baylor University,
Waco, Texas, U.S.A.*

**KLUWER ACADEMIC PUBLISHERS**
DORDRECHT / BOSTON / LONDON

Library of Congress Cataloging-in-Publication Data

```
Toombs. S. Kay.. 1943-
    The meaning of illness : a phenomenological account of the
different perspectives of physician and patient / by S. Kay Toombs.
        p.  cm. -- (Philosophy and medicine ; v. 42)
    Includes bibliographical references and index.
    ISBN 0-7923-1570-7 (HB : alk. paper)
    1. Sick--Psychology. 2. Physicians--Psychology. 3. Medicine and
psychology. 4. Physician and patient. 5. Phenomenological
psychology.  I. Title.  II. Series.
    [DNLM. 1. Disease--psychology. 2. Physician-Patient Relations.
W3 PH609 v.42 / W 62 T672m]
R726.5.T655  1992
610.89'6--dc20
DNLM/DLC
for Library of Congress                                    91-46512
```

ISBN 0-7923-1570-7

Published by Kluwer Academic Publishers,
P.O. Box 17, 3300 AA Dordrecht, The Netherlands.

Kluwer Academic Publishers incorporates
the publishing programmes of
D. Reidel, Martinus Nijhoff, Dr W. Junk and MTP Press.

Sold and distributed in the U.S.A. and Canada
by Kluwer Academic Publishers,
101 Philip Drive, Norwell, MA 02061, U.S.A.

In all other countries, sold and distributed
by Kluwer Academic Publishers Group,
P.O. Box 322, 3300 AH Dordrecht, The Netherlands.

*Printed on acid-free paper*

02-0498-200 ts

Printed in the Netherlands

*For my mother and father,
Alan and Kathleen Bell,
and my husband, Dee*

# CONTENTS

# ACKNOWLEDGMENTS

I should like to express my deep appreciation to the following individuals who have provided assistance, inspiration and encouragement throughout various stages of the development of this work.

Richard Zaner, Edmund Pellegrino, Eric Cassell and Richard Baron first convinced me that it was important to develop my ideas and to do so in a philosophically rigorous fashion. In sharing their work and in generously responding to my ideas, they provided the impetus for what has become an intensely challenging and stimulating lifelong project. I will always be grateful to them. In addition, Richard Zaner contributed directly to this manuscript by reading it and providing invaluable criticisms.

Steven Crowell is in large part responsible for providing me with the intellectual tools to carry out my project. He nurtured my interest in phenomenology and shared my excitement as I discovered new insights. I could not have asked for a more responsive teacher.

H. Tristram Engelhardt, Jr. has not only provided helpful critical comments but he has invariably caused me to reflect more deeply and to discover new and interesting ideas. Our exchanges have been both stimulating and rewarding.

Stuart Spicker has contributed greatly to the development of this manuscript. His careful textual criticism has prompted me to refine the work in important ways. While he is in no way responsible for what is finally said here, I warmly acknowledge his assistance.

One of the most rewarding aspects of this endeavor has been the participation of the following individuals who have taken time out of their busy lives to make comments and express ongoing support for my work. Their encouragement has been a highly motivating factor throughout the course of this project: Jerome Bruner, Jay Katz, Ian McWhinney, Michael Schwartz, and Oliver Sacks. Special thanks are due to Charles Silberman who has graciously put aside his own writing to read mine and to make detailed and insightful suggestions. Furthermore, at those times when I doubted my ability to complete the task, he assured me that the end result would be worth the effort and he motivated me to continue.

There is one person without whom this book would never have been written. My husband, Dee Toombs, is the source of whatever strength I have and the one constant in a life that is necessarily filled with uncertainty. His unwavering love and support gives me the freedom and the courage to pursue those goals and projects I hold dear, and to continue to do so in the face of progressive physical disability. In this way he keeps me whole.

I would also like to thank Kluwer Academic Publishers for permission to include material which was previously published in the following articles: Toombs, S.K.: 1988, "Illness and the paradigm of lived body." *Theoretical Medicine* 9 (June 1988): 201–26; Toombs, S.K.: 1987, "The meaning of illness: A phenomenological approach to the patient-physician relationship." *Journal of Medicine and Philosophy* 12 (August 1987): 219–40; Toombs, S.K.: 1990, "The temporality of illness: Four levels of experience." *Theoretical Medicine* 11, 227–241.

# A PHENOMENOLOGICAL APPROACH

My interest in exploring the nature of the patient's and the physician's understanding of illness has grown out of my own experience as a multiple sclerosis patient. In discussing my illness with physicians, it has often seemed to me that we have been somehow talking at cross purposes, discussing different things, never quite reaching one another. This inability to communicate does not, for the most part, result from inattentiveness or insensitivity but from a fundamental disagreement about the nature of illness. Rather than representing a shared reality between us, illness represents two quite distinct realities – the meaning of one being significantly and distinctively different from the meaning of the other.

In this work I shall suggest that psychological phenomenology provides the means to examine the nature of this fundamental disagreement between physician and patient in a rigorous fashion.[1] In particular, psychological phenomenology discloses the manner in which the individual constitutes the meaning of his or her experience.

In providing a phenomenological description,[2] the phenomenologist is committed to the effort to begin with what is given in immediate experience, to turn to the essential features of what presents itself as it presents itself to consciousness, and thereby to clarify the constitutive activity of consciousness and the sense-structure of experiencing.

One of the primary aims of an explicitly phenomenological approach is to let what is given appear as pure phenomenon (the thing-as-meant) and to work to describe the invariant features of such phenomena. As Edward Casey (1976, pp. 8–9) notes "the phenomenologist's basic attitude is: no matter how something came to be in the first place, what is of crucial concern is the detailed description of the phenomenon *as it now appears*."

A phenomenological approach thus involves a type of radical disengagement, a "distancing" from our immediate ongoing experience of everyday life in order to make explicit the nature of such experience and the essential intentional structures which determine the meaning of such experience.[3] As such, phenomenology is an essentially reflective enterprise. The common sense world itself (and our experiencing of it) becomes the focus of our reflection. As Richard Zaner (1970, p. 51)

xi

points out, our attention shifts from that of engagement in the world to
that of focal concern for the sense and strata of the very engagement
itself. Rather than straightforward, unreflective absorption in the objects
of experience, the phenomenological approach involves reflection *upon*
experience. The task is to elucidate and render explicit the taken-for-
granted assumptions of everyday life and, particularly, to bring to the fore
one's consciousness-of the world. In rendering explicit the intentional
structures of consciousness, phenomenological reflection thematizes the
meaning of experience.[4]

In order to describe phenomena as they present themselves to consciousness, the phenomenologist attempts to effect a systematic
neutrality. That is, in the phenomenological attitude, one places in
abeyance one's taken-for-granted presuppositions about the nature of
"reality," one's commitments to certain habitual ways of interpreting the
world. In particular, the phenomenologist sets aside any theoretical
commitments derived from the natural sciences (for example, the causal-
genetic mode of analysis) in order to describe what gives itself directly to
consciousness.[5]

As Maurice Natanson explains with regard to Edmund Husserl's notion
of bracketing or phenomenological reduction, what is thus disclosed in
the reduction is *the field of intentionality* – the conscious processes of
experiencing ("the noetic") and the objects of experience ("the
noematic").

To bracket the world is neither to deny its reality nor to change its reality in any way;
rather, it is to effect a change in my way of regarding the world, a change that turns
my glance from the "real" object to the object as I take it, treat it, interpret it as real.
Within the natural attitude I attend to the object; in the phenomenological attitude I
attend to the object as known, as meant, as intended.... The object continues to be in
the real world as I do, but what now interests me, phenomenologically, is my
awareness, my sense of its being in the real world. The object I reflect upon in the
reduced sphere is the real thing as I've taken it to be real. Thus, "the" world is
replaced by "my" world, not in any solipsistic sense, but only in the sense that "mine"
indicates an intentional realm constituted by my own acts of seeing, hearing,
remembering, imagining, and so on (Natanson, 1968, pp. 58–59).

This "radical" reflection does not deny the existence of the physical,
social and cultural world but rather reveals the "prejudices" and taken-
for-granted presuppositions which are not explicitly recognized in our
spontaneous, unreflective experience. Indeed, the phenomenological

reduction alone discloses the "setting of the world" which is presupposed at every moment of our thought.

The philosopher, in so far as he is a philosopher, ought not to think like the external man, the psychophysical subject who is *in* time, *in* space, *in* society, as an object is in a container. From the mere fact that he desires not only to exist but to exist with an understanding of what he does, it follows that he must suspend the affirmations which are implied in the given facts of his life. But to suspend them is not to deny them and even less to deny the link which binds us to the physical, social and cultural world. It is on the contrary to *see* this link, to become conscious of it (Merleau-Ponty, 1962, pp. vii-xxi).[6]

With its emphasis on firsthand or direct description, phenomenology provides the means to elucidate the domain of unreflective, taken-for-granted lived experience, to provide a detailed account of the manner in which we interpret the world of everyday life (the lifeworld).[7] In particular, phenomenology discloses that meaning is constituted in light of certain invariant intentional structures which characterize consciousness. Such structures should be recognized if one is to give an account of meaning. Furthermore, phenomenology provides an explication of the fundamental and important distinction between lifeworld interpretation and scientific conceptualization.

In sum, then, the psychological phenomenological approach includes the following: (1) the effort to elucidate the manner in which meaning is constituted; (2) the commitment to a radical reflection upon lived experience which requires (as a methodological device) the setting aside of theoretical commitments and taken-for-granted common-sense presuppositions in order to focus upon the conscious processes of experiencing and the objects of experience; and (3) the attempt to uncover the invariant features of phenomena and thereby to provide a rigorous description of such phenomena.[8]

This approach provides a powerful means to render explicit the different perspectives of physician and patient. In particular, such an approach focuses explicitly upon the phenomenon of illness and the manner in which meaning is constituted by the patient and the physician.

It must be emphasized that the psychological phenomenological analysis of the constitution of meaning is in no way to be equated with empirical psychology.[9] To do so is to misunderstand the nature of phenomenological method – particularly the phenomenological reduction which involves the setting aside of all the presuppositions concerning the nature of the world which are operative at the level of empirical science. That is, empirical science begins with (and remains within) the "natural attitude," whereas psychological phenomenology involves a radical disengagement from the "natural attitude" in order to analyze critically the nature of experiencing as such.[10] Thus, whereas empirical psychology is concerned with psychological "facts," psychological phenomenology is concerned to describe the objectivating activity of human consciousness and the manner in which meaning is constituted in everyday life.[11]

In this regard, psychological phenomenological description focuses on the phenomenon of illness, as it is constituted at different levels of meaning (both pre-reflective and reflective). Such an analysis discloses not only the enormous complexity of the meaning of illness, but the fundamental distinction between meaning which is grounded in lived experience and meaning which represents an abstraction from lived experience. Furthermore, in focusing on the phenomenon of illness, the task is to elucidate invariant features of illness-as-experienced apart from the varieties of its concrete instantiations.

In order to make explicit the different perspectives of physician and patient with regard to the meaning of illness, certain issues have to be addressed. One of these is the structure of intersubjective agreement – how is successful communication with another Self possible given that my experience is essentially "mine"? How is a shared world of meaning constituted in everyday life?[12] What light does the analysis of intersubjectivity at the most general level shed on the problem of intersubjective agreement in the patient-physician encounter?

A further issue arises concerning the distinction between the immediate lived experience of illness and the conceptualization of illness as a disease state. What is the nature of this distinction? How do physician and patient experience illness differently?[13]

Further, since physical illness involves the body, an exploration into the difference in understanding between physician and patient must explore the manner in which the body is apprehended by each. In particular, in order to understand the nature of illness and the experience of the patient, it is necessary to focus upon the lived experience of embodiment. How does such experience manifest itself in "normal" circumstances? In illness? How does the patient's apprehension of the body-in-illness differ from the physician's conception of the diseased body?[14]

In what follows I shall suggest that psychological phenomenology provides important insights into these questions and, thus, into the differing and unique perspectives of physician and patient. In particular, it will be noted that these different perspectives are emblematic of a systematic distortion of meaning which emerges in the patient-physician encounter. This distortion of meaning relates not only to the nature of illness but also to the nature of the experience of the body. Rather than communicating on the basis of a shared set of assumptions (i.e., perceptions), physician and patient routinely communicate with one another from within the context of different "worlds," each "world" providing a unique horizon of meaning.

In clarifying the different perspectives of physician and patient, I shall show that this is not simply a matter of different levels of knowledge (too often assumed to be the case) but that the difference in understanding is much more profound. In particular, I shall note that it is imperative to differentiate between meaning which is grounded in immediate experience and meaning which is not so grounded. Following Husserl, I shall make a distinction between the "natural" attitude which involves the immediate pre-theoretical experiencing of the world of everyday life, and the "naturalistic" attitude which involves an essential abstraction from immediate experiencing in favor of a theoretical, scientific account of the causal structure of such experiencing. This distinction is particularly important in the physician-patient relationship where the decisive gap between lived experience and the scientific account of such experience clash in a direct way with regard to the phenomenon of illness.

The decisive gap between lived experience and scientific explanation (which is disclosed in the phenomenological analysis) is at the root of the fundamental distortion of meaning in the physician-patient relationship – yet it is not generally explicitly recognized, nor is its impact on the patient-physician relationship well understood.[15]

I shall argue that, in fact, it has a major impact on the relationship. In particular, it will be shown that the phenomenon of illness-as-lived is quite distinct from the phenomenon of the disease state and that the two cannot be identified with one another. Thus, when physician and patient talk about "illness" they are not discussing a shared "reality."

An explicit recognition of this fundamental distinction between illness and the disease state is important for developing shared meaning between doctor and patient. It also has significant implications for medical practice (implications which will be discussed in detail throughout the course of the analysis). For example, since illness-as-lived is quite distinct from the phenomenon of the disease state, it is necessary that the physician explicitly attend to the lived experience when devising therapeutic goals. In this regard the phenomenological analysis of the body provides the insight that illness is fundamentally experienced as the disruption of lived body rather than as the dysfunction of biological body. Thus, if therapeutic goals are to be optimally effective – and suffering is to be relieved – attention must be directed to the patient's perceived lived body disruption rather than being exclusively directed towards the objective pathophysiology of the disease state.

The phenomenological analysis carried out in this work indicates that the prevailing biomedical model of disease (which tends to focus exclusively on the dysfunction of the biological organism and the pathophysiology of the disease state) is an incomplete model for medical care. It will be suggested that an adequate account of illness must include not only a construal of illness in terms of clinically definable disease states but also an understanding of illness-as-lived. In this regard it will be noted that illness may be understood as a particular way of being in the world – a way of being that exhibits certain typical characteristics. Such characteristics must be recognized if one is to grasp what illness means to patients.

Although the major emphasis of this work is to explicate the different perspectives of physician and patient, I shall argue that the phenomenological analysis also provides clues as to the manner in which a shared world of meaning may be constituted between them. It will be noted that the lifeworlds of physician and patient provide the starting point for mutual understanding of the illness experience. In particular, I shall argue that reflection upon the manner in which the body is apprehended in everyday life can provide the basis for empathic understanding of the "givenness" of illness. In addition, clinical narratives provide insights into the lived experience of illness.

Finally, it will be noted that the physician-patient relationship is a unique kind of "face-to-face" relationship in that it is grounded in the patient's experience of illness. A distinction will be made between "healing" and "curing" disease and it will be argued that "healing" requires an understanding of illness-as-lived.

# CHAPTER ONE

# THE SEPARATE WORLDS OF
# PHYSICIAN AND PATIENT

## 1. OWN WORLD[16]

In his descriptive investigation of phenomena, Husserl paid particular attention to the manner in which the individual experiences the world. He analyzed such experiencing in terms of the structuring activity of consciousness, and thereby disclosed an essential correlation between the perceiver and the object perceived (e.g., between myself-as-believing and the belief-as-believed-by-me). In disclosing this essential correlation Husserl demonstrated that immediate experiencing is necessarily unique.[17]

In emphasizing the direct exploration of experienced phenomena, Husserl was concerned that we critically evaluate all our presuppositions about the everyday world (the lifeworld). He noted that we tend not to reflect consciously upon the manner in which we experience "reality"; we simply take the "objectivity" of the everyday world for granted, rarely taking it to be a world constituted by the continuous activity of consciousness.[18] Husserl (1962, pp. 96 ff.) called for the phenomenologist to suspend this unreflective, taken-for-granted acceptance of the everyday world (a cognitive act he referred to as the phenomenological "reduction" or "bracketing"). In performing the phenomenological reduction, the activity of experiencing itself becomes explicit. The phenomenologist's concern is no longer with the object-as-such, but rather with the object-as-perceived, as-experienced.

In the course of his analysis of experiencing, Husserl identified certain essential features (intentional structures) which characterize consciousness. It is in light of such intentional structures that one may provide a rigorous analysis of the manner in which meaning is constituted.

### The Theory of Intentionality

According to Husserl the fundamental feature of consciousness is

1

intentionality. All consciousness is necessarily a consciousness-of. Consequently, one cannot understand consciousness without referring both to *the act* of consciousness and *the intended object* of consciousness. Consciousness is directional in the sense that to be conscious-of is to be directed toward an object (i.e., the nature of acts of consciousness is such that they point toward their intended objects). All thinking is thinking of something, all perceiving is perceiving of something, all imagining is imagining of something. Thus, there is an *essential correlation* between acts of consciousness (e.g., perceiving, remembering, imagining, and so forth) and the intended objects of consciousness, such that objects of consciousness are to be understood not as "things" but rather as correlates of intentional acts. For example, I glance out of my window and see (perceive) the tree to the right of the garden furniture. Later I recall this same tree in an act of remembering, or perhaps engage my imagination in contemplating how the tree might appear. In each act the *intentional* object, "tree-as-intended," derives its *sense* from my intending act. In the first case the intentional object is the tree-as-perceived; in the others it is tree-as-remembered or tree-as-imagined. To consider the manner in which the meaning of intended objects is constituted by consciousness is, therefore, necessarily to take into account the interrelated complex of experiencing – the intentional acts (in Husserl's terms "noeses") and the intended objects ("noemata").

## *Directionality (Focusing)*

Husserl thus noted that the manner in which an object is intended is strictly correlative to the way in which consciousness explicitly attends to, or directs itself to, that object. In Husserl's (1970b, p. 108) terms the activity of consciousness renders the intended object "thematic." It is through such attentional focusing that certain aspects of the object are rendered explicit. One may, for example, attend to the color rather than the taste of a glass of wine. One may attend to Elizabeth II *as* Queen of England, or *as* wife of Philip. The attentional focus that renders the intended object thematic may vary. Additionally, one may attend in a variety of modes – cognitively, valuationally, emotively, and so forth (Zaner, 1970, p. 165). The meaning of the object-as-experienced will thereby change as the attentional focus varies.

## *Temporality*

Husserl (1962, pp. 91–93, 215–20) notes that a fundamental feature of consciousness is temporality. Not only is it the case that each particular consciousness-of something exhibits a temporal structure but temporality is the unitary form binding all experiences within a single "stream of experience."[19] The temporal aspects of the noetic-noematic structure are crucial in the constitution of the object-as-meant and especially in the constitution of intersubjectivity. In this connection, Husserl (1982, pp. 39–43) distinguishes between the consciousness of internal time (an aspect of the noetic structure) which synthesizes the intended object as a coherent whole, and its correlate, immanent temporality (an aspect of the noematic structure) by means of which whatever is experienced is temporally ordered.

Experiencing exhibits a horizonal structure. This horizonal structure is temporal as well as spatial. The ongoing stream of consciousness is such that every "now" perception is a temporal phase in a continuously flowing succession of present/just past/new "now" moments of perceiving. Nevertheless, objects are experienced as temporal unities. The "now" perception appears not as a discrete, isolated instant along a given time-line but rather as an integral part of a continuum – a continuum which incorporates not only the present now-point but those now-points which are just past, as well as future now-points which are to come. Every present moment of experience has about it a "fringe" of experiences (a moment just-past and a successive future moment) which are a part of the present consciousness of the object (Husserl, 1964; 1962, pp. 218–219). In the constitution of this temporal unity, Husserl identifies a particular kind of memory – primary memory ("retention"). The enduring consciousness of the object is such that past temporal phases of the object are retained in primary memory as a part of the present consciousness of the object. Furthermore, future phases are protended (or anticipated) in the now-consciousness. David Carr (1986, pp. 23–24) notes that Husserl's analysis of internal time-consciousness provides the key insight that the temporal must be considered as a "field of occurrence" with past and future providing the horizons for the present. Temporal consciousness can be compared to "a gaze which spans or takes in the temporal horizons of future and past, against which the temporal object presents itself."

Husserl's analysis of internal time-consciousness reveals a radical distinction between lived time and objective time. Lived time is the ongoing, immediate experiencing of the temporal phases of an object through the interplay of retentions and protentions which are evoked in the stream of consciousness. Objective time, on the other hand, is the time that can be measured by clocks, calendars, and so forth. Alfred Schutz (1976a, pp. 159–178) notes that the distinction between lived and objective time is readily apparent when considering, for example, the experience of a person listening to music. While living through the ongoing flow of the music, the listener is not aware of objective time. It may come as a complete surprise that one movement in the music takes as much time (in the clock sense) as another movement. While actually experiencing the music, the listener is immersed in its ongoing flow, in the ongoing articulation of the musical piece. In living through the ongoing flow of internal-time consciousness, the individual "lives in" a dimension of time which is incomparable with that which can be measured according to the objective time scale. Other experiences are equally indicative of the incommensurability of lived and objective time. "The hand of our watch may run equally over half the dial, whether we wait before the door of a surgeon operating on a person dear to us or whether we are having a good time in congenial company" (Schutz, 1976a, p. 171).

*Horizon*

The horizonal structure of experiencing is not only temporal but spatial (Husserl, 1962, pp. 91–93). Everything we encounter is encountered as a "being-in-a-context" (Zaner, 1970, p. 154). For example, when I turn to look at the tree in my garden, I perceive the tree as outside my window, next to the garden furniture, in front of the fence, and so forth. The tree-as-seen is set off from a background of co-perceived things (things which can themselves become the explicit focus of my attention if I choose to direct my attention to them).

In addition, my perception of the tree includes aspects present in the perceived, but not yet actually perceived. For example, my vantage point from the study allows me to view only one particular side of the trunk and prevents me from seeing the top branches. Nevertheless, the intentional object (tree-as-seen) is interpreted by me as a coherent whole (i.e., I interpret the tree as an object having a back side and a top. The perceived aspects of the tree point to those aspects which are not explicitly the focus of my attention). Moreover, I am aware that if I changed position I could bring these presently unseen aspects of the tree into focus (I could walk around the tree to the back side and I could even perceive the top branches of the tree if I climbed on the roof of my house or flew over the tree in a helicopter). Additionally, my present perception is connected to previously experienced perceptions of this particular tree (as it appeared last spring, as it looked when I walked around it yesterday, and so forth) and to my experience of trees in general.

Furthermore, the horizon is not only spatial and temporal but also social, historical, economic, political, and so on (Zaner, 1970, p. 154). No object is perceived as insulated but rather each is comprehended as an object within a "horizon of familiarity and preacquaintanceship" (Schutz, 1962b, p. 7).

The field of consciousness is, thus, structured into a thematic kernel which stands out over against a surrounding horizon. The horizon is constituted not only by perceptual experiences (e.g., the background of co-perceived things) but also by one's former experiences which are preserved in memory or available within one's present stock of knowledge. Thus, the meaning of a particular object cannot be separated from the global field of meaning of the individual's world.

### *Biographical Situation*[20]

In reflecting upon the constitution of meaning in everyday life, it is important to note the extent to which the individual's unique biographical situation determines the manner in which intended objects are thematized; that is, just what the individual attends to depends upon his or her situation within the social world and upon the complicated texture of choices, decisions and projects that make up a life plan. Experience is encountered, attended to, and rendered thematic in terms of one's unique situation, in light of one's own special interests, motives, desires, aspirations, religious and ideological commitments, and so forth.[21]

Each individual occupies a unique biographical situation within the social world. This situation is unique not only in terms of one's actual physical environment, but also in the manner in which one arrives at a carefully constructed definition of reality (prior to adopting the phenomenological attitude). At the outset we interpret common sense reality in light of certain naive typifications. For example, when I see my neighbor waving his hand, I typify this behavior as a friendly greeting, just as I typify the object outside my window as a tree, and the object moving along the street as a car. Such typifications are, for the most part, culturally and socially inculcated and they bestow on everyday life its quality of taken-for-grantedness. Nevertheless, it should be noted that from the outset such typifications are handed down from parents and teachers in such a way that our interpretation of the surrounding world carries with it a "sedimented" meaning which is different for each one of us.[22] Additionally, as we live and act in everyday life, we amass a store of subjective experiences and compile a unique stock of knowledge which is necessarily ours alone and upon which we build further interpretations of reality.

In this regard, it is important to note that the horizonal temporal structure of experiencing (which has been explicated above) is evident in the constitution of meaning in the individual's biographical situation. As Schutz (1962f, p. 307) notes, the world is organized around the individual as center not only spatially but temporally (i.e., my actual "Now" is the origin of all time perspectives – past and future, sooner and later, and so forth). Most importantly, Schutz points out that the actual present (a phase or element of the individual's unique biographical situation) transcends the Here and Now, in that the present incorporates the individual's recollections of the way things have appeared in the past and anticipations of the way they will appear in the future.[23]

Not only is this horizonal temporal structure evident in such constitution of meaning but Husserl's analysis of time provides the key insight that this temporal structure pervades lived experience. The present is not an isolated instant along a given time-line but rather a present-now which is always experienced within the horizons of past and future. Carr (1986, pp. 18–72) makes the point that this temporal structure is evident at all levels of experiencing, from the most fundamental level of pre-reflective sensory experience, through the level of simple actions, up to the level of complex sequences of action. Such horizonal temporal structure is also exhibited at the level of the life narrative (Carr, 1986, pp. 73–99; MacIntyre, 1981, pp. 190–209).[24] The present is interpreted in light of past experiences and future anticipations.

## Summary

In summary then, as individuals, we find ourselves always located within the everyday world, the world of immediate experience.[25] In order to render this world comprehensible, we interpret it in light of a meaningful structure which we impose upon the "reality" we encounter. By means of intentionality and directedness of acts of consciousness, we attend to certain aspects of our experience which are always perceived against a background or horizon which includes a unique biographical situation and stock of knowledge, and which incorporates the horizons of past and future.

Since all experience represents a correlation between experiencing and that which is experienced, and since the locus of meaning is grounded in the intentional activity of consciousness, the Other's experiencing is not immediately accessible. Everyone has exclusively his or her own phenomena which only he/she is capable of experiencing quite originally (Bosch, 1970, p. 55). As Husserl (1982, pp. 113 ff.) notes the contents of another's world are, therefore, only available to me in an "appresent" manner. I cannot experience them directly.[26]

Each individual retains the essential core of experiencing as a constituted world. In such a "world-for-me" things are not "the in-themselves-existing things of nature – of the exact sciences with the definitive properties which alone are recognized by science as objective characteristics – instead they are experienced, thought, or otherwise posited things as such, intentional objectivities of the personal consciousness" (Bosch, 1970, p. 54).

Nevertheless, although, as individuals, we experience the world directly in a unique way, we perceive ourselves to be located in an intersubjective world, that is, to be living in the familiar world among fellow men who share a relationship to a common world. I perceive myself to be an experiencing subject (for whom objects exist as correlates of my experiencing of them) among other experiencing subjects for whom I, myself, exist as an object. Thus, I am at once subject and object in the world. In addition, I am a self-conscious being in that I am reflectively conscious of being both subject and object in an intersubjective world.[27]

How is the intersubjective world constituted? And, particularly, how is a shared relationship to a common world possible given the unique nature of experiencing?

## 2. COMMON WORLD

Schutz notes that the world of everyday life is from the outset an intersubjective world.[28] The individual is located within a historical, social, cultural environment, as a man among other men. Even one's unique biographical situation is *to some extent* a shared situation. That is, one's stock of knowledge (the set of typifications one uses to interpret mundane reality) is for the most part culturally and socially derived. In the "natural attitude" the individual takes for granted "the bodily existence of other men, their conscious life, the possibility of intercommunication, and the historical givenness of social organization and culture" (Schutz, 1962f, p. 313).

Nevertheless, since all experiencing is necessarily unique, one can have only indirect knowledge of another's experience – such indirect knowledge being available through events in the outer world such as the Other's bodily gestures and linguistic expressions.[29] In other words, while it is the case that I perceive the Other's body directly, I can have no immediate apprehension of his or her thoughts or experience. It is through the medium of bodily events (e.g., blushing, smiling), body movements (e.g., wincing, beckoning), bodily activities (e.g., walking, talking), as well as communication through language, that I interpret the Other's thoughts and experience (Schutz, 1962f, p. 315). Such forms of appresentational reference (bodily expressions and the communicative process) have the function of establishing a common environment (a shared world of meaning) – at least to some extent, although fully successful communication remains unattainable in principle given the unique nature of experience.[30] There will always remain an inaccessible zone of the Other's private world which transcends my possible experience.[31]

Schutz notes that certain idealizations – the idealization of the interchangeability of standpoints and the idealization of the congruency of the system of relevances – are presupposed if a shared world of meaning (a "communicative common environment") is to be established. That is, a world of "common" objects is made possible by such idealizations and they are thus the basis for communication (Schutz, 1962b, pp. 11–12). In the ordinary course of events the "same" object means something different to myself and to my fellows in that each of us experiences the object from a different perspective spatially and in light of unique biographical situations, different purposes at hand and different systems of relevances. Common sense thinking overcomes these differences in individual perspective by means of two typifying constructs which Schutz has called "the general thesis of reciprocal perspectives." This general thesis is comprised of two idealizations:

(i) The idealization of the interchangeability of the standpoints: I take for granted – and assume my fellow-man does the same – that if I change places with him so that his "here" becomes mine, I shall be at the same distance from things and see them with the same typicality as he actually does; moreover, the same things would be in my reach which are actually in his. (The reverse is also true.)

(ii) The idealization of the congruency of the system of relevances: Until counter evidence I take it for granted – and assume my fellow-man does the same – that the difference in perspectives originating in our unique biographical situations are irrelevant for the purpose at hand of either of us and that he and I, that "We" assume that both of us have selected and interpreted the actually or potentially common objects and their features in an identical manner or at least an "empirically identical" manner, i.e., one sufficient for all practical purposes (Schutz, 1962b, pp. 11–12).

Both idealizations are "typifying constructs of objects of thought which supersede the thought objects of my and my fellow-man's private experience." It is largely due to the operation of these constructs of common-sense thinking that the *private* world of immediate experiencing is rendered into a *common* world shared with other fellow men.

### Summary

In sum, then, the common world is constituted in the encounter with other individuals through the establishment of a "communicative common environment." Such a "communicative common environment" is possible because the familiar world is interpreted by means of typifications, a stock of knowledge which is socially and culturally derived, and in light of the "general thesis of reciprocal perspectives." Successful communication thus presupposes a certain taken-for-granted congruence in the appresentational and interpretational schemes of the communicators (Schutz, 1962f, p. 327). As Schutz notes with regard to communication:

(1) since the sign used in communication is always preinterpreted by the communicator in terms of its expected interpretation by the addressee, communication presupposes that the interpretational scheme which the communicator relates and that which the interpreter will relate to the communicative sign in question will *substantially* coincide (Schutz, 1962f, p. 322);

(2) full identity of the interpretational schemes of communicator and interpreter is impossible (because such interpretational schemes are determined by the unique biographical situations of each of them). Successful communication is, therefore, possible only between persons who share a substantially similar system of relevances. The greater the differences between the systems of relevances, the fewer are the chances for successful communication (Schutz, 1962f, pp. 322–323);

(3) to be successful the communicative process must involve a set of common abstractions or standardizations. Typification is a form of abstraction which provides the basis for standardization within common-sense thinking. Typification occurs within the prepredicative sphere of experience (i.e., prepredicative experience is organized from the outset under certain types) (Schutz, 1962f, p. 323).

### 3. DIFFERENT PERSPECTIVES OF PHYSICIAN AND PATIENT

The foregoing phenomenological analysis of "own world" and "common world" provides important insights into the constitution of meaning in the context of the physician-patient relationship. Such an analysis reveals that there is a systematic distortion of meaning in the physician-patient relationship. In particular, illness is experienced in significantly different ways by physician and patient. Consequently, rather than representing a shared "reality" between them, illness represents in effect two quite distinct "realities."

A consideration of such phenomenological concepts as focusing (directionality), the naturalistic attitude, temporality, and relevance reveals that the physician and patient apprehend illness from within the context of separate worlds, each world providing its own horizon of meaning. Furthermore, it becomes clear that the experience of illness is such that it is particularly difficult to construct a shared world of meaning between physician and patient.

### *Focusing*

As Husserl has noted, the manner in which an object is experienced is strictly correlative to the way in which an individual attends to it. The activity of consciousness renders the object "thematic." Such attentional focusing determines the meaning of illness. The patient and physician are motivated to attend to different aspects of the experience, and each thereby renders it thematic in a qualitatively distinct manner. The physician is trained to perceive illness essentially as a collection of physical signs and symptoms which define a particular disease state. He or she thematizes the illness as being *a particular case* of "multiple sclerosis," "diabetes," "peptic ulcer," and so forth. The patient, however, focuses on a different "reality." One does not "see" one's own illness primarily as a disease process. Rather, one experiences it essentially in terms of its effects upon everyday life. Thus, whereas the physician sees the patient's illness as a typical example of a disease, the patient attends to the illness for its own sake. This is an explicitly different focus. Whenever one considers something as an example, it is not considered for its own sake, but only insofar as it exemplifies something other than the affair itself.[32]

The motivation for focusing is intimately related to the individual's placement within the familiar world. In the practice of a profession certain "habits of mind" develop that provide a horizon of meaning by means of which reality is interpreted. Such "habits of mind" are in many ways peculiar to the profession that utilizes them. They represent a distinct approach to the world and compose the culture of a profession (Kestenbaum, 1982a, pp. 6–7).

"Habits of mind" in a real way determine the manner in which an object is rendered thematic. For example, the professional art critic and the ordinary "man-in-the-street" will look at a painting differently. The art critic will be influenced by certain "habits of mind" that are a function of the art profession. He or she may be preoccupied with the technique of the artist, the explicit use of color, and so forth. These "habits of mind" will, to a large extent, determine what is "seen" and the way in which the sense of the object is made explicit. The art critic's experiencing will, therefore, be quite different from that of the untrained individual. Indeed, it may be difficult for them to converse together about the same painting in anything other than a very superficial manner.

The scientific "habit of mind" likewise determines the manner in which an object is rendered thematic. It provides a horizon of meaning, a motivation for focusing, and a means of interpreting "reality." However, the scientific interpretation is quite distinct from other interpretations of "reality." In particular, it is quite different from the immediate experiencing of that "reality" in the everyday world.[33]

Natanson argues that the world of immediate experience has a certain precedence over the derivative world of science. We first of all experience the "world" in its immediacy. Only in reflection and abstraction may we then thematize our experience in terms of theoretical, scientific constructs. Even then, as Natanson (1968, p. 95) notes, there is a "decisive gap" between one's immediate experiencing and the theoretical, scientific account of the causal structure of such experiencing.

A study of Helmholtz's Physiological Optics tells me nothing about the visual experience I have in its qualitative immediacy ... my color world is first of all mine; it is not mediated by expert knowledge of its conditions, nor is the theory of vision in any way relevant to its presentational validity. It is only in a derivative sense that the case of my color experience falls under the general scientific category of visual perception. In one sense, then, my color world is a privileged one: the total scope and content given in it possess an experiential depth that is independent of subsequent theoretical explanation (Natanson, 1968, p. 95).

Natanson points out that what holds for vision holds for the entire realm of immediate experience.

This "decisive gap" between immediate experience and the derivative realm of science manifests itself concretely in the experience of illness. Patients encounter illness in its qualitative immediacy. The categories that they use to define illness are primarily concerned with everyday life and functioning.[34] Physicians, on the other hand, may categorize the patient's illness solely in terms of scientific constructs; that is, according to the prevailing "habits of mind" of the medical profession that render the illness thematic in terms of "objective," quantifiable data. Indeed, it is often assumed by physicians that such clinical data *exclusively* represent the "reality" of a patient's illness.[35] As Eric Cassell (1983) notes, on being presented with a sick person doctors do not attempt to find out what is the matter but, rather, attempt to make a diagnosis. This is not the same thing. Diagnoses are "relatively sharply defined name diseases that are believed to exist when certain criteria are met by the patient's history, physical examination, or laboratory or other tests" (Cassell, 1983). In the event that such objective criteria are not met, it may be concluded by the physician that the patient's complaint is not a *bona fide* illness. Nevertheless the patient still experiences him or herself as sick.

It is worth noting that when physicians themselves become patients they immediately become aware of the "decisive gap" between the qualitative immediacy of their own experience of illness and any subsequent scientific explanation in terms of disease (Sacks, 1984). "I practiced medicine for 50 years before I became a patient," says Dr. Edward E. Rosenbaum, former chief of rheumatology at Oregon Health Sciences University. "It wasn't until then that I learned the physician and patient are not on the same track. The view is entirely different when you are standing at the side of the bed from when you are lying in it" (Rosenbaum, 1988, p. viii). Physicians who speak of their experience as patients not only note this change in experiencing but say they have great difficulty discussing their illness with colleagues (Rabin, 1982, pp. 506–509; Lear, 1980; Stetten, 1981; Mandell and Spiro, 1987). What they explicitly fail to recognize is the difference in thematizing. Their colleagues are thematizing the illness according to the "habits of mind" of the profession, whereas they, as patients, are responding to the illness-as-lived.

## The Natural and the Naturalistic Attitude

The foregoing distinction between immediate experience and the derivative world of science has been further explicated by Husserl (1970, pp. 321, 379) in terms of the distinction between the "natural attitude" and the "naturalistic attitude." In the "natural attitude" the world itself is not made explicitly thematic as an intended object. Rather we ordinarily take the existence of the world (and the objects within it) for granted and act in the world in a pragmatic fashion according to our subjective and selective interests.

That is, we find ourselves always within the world of immediate experience. It is presupposed and pregiven in all that we do and we take its validity for granted without explicitly investigating the world *as* world. The "natural attitude" is prior to all scientific intent and activity.

As Kohak explains:

[W]hen Husserl speaks of the natural standpoint, he is not claiming that some biological necessity forces a metaphysics upon us but simply that the ingrained habits of our common sense, without our even being aware of it, lead us into a metaphysics. Precisely because our common sense is habitual and quite free of self-conscious reflection, because it is preoccupied with the world it encounters as "out there," it assumes that reality itself is "out there," only passively recorded by the subject "in here." The shift is subtle but significant. As lived, reality is the experiencing of an object. As common sense interprets it, the reality is the object, the experience is incidental to it. That is no longer a datum; it is the unacknowledged theoretical postulate of common-sense knowledge. Husserl calls it the "thesis of the natural standpoint": the world is "out there," only its reflection is "in here." I must look for an explanation "out there"; or in sum, lived experience is what is to be explained, and the world is what explains it. To understand my experience, common sense assumes, I need to know what I am experiencing but must discover what in the world is causing it (Kohak, 1978, p. 32).

In the "naturalistic attitude," however, the intent is to thematize the world as "object" and to consider the world itself as a scientific theme. The aim in the "naturalistic" (or scientific) attitude is to grasp the nature of "reality" and to describe such "reality" in terms of some "objective" description which will accurately characterize the "thing-in-itself" apart from one's experiencing of it.

As Natanson has indicated in the above distinction between the world of immediate experience and the world of science, there is a fundamental difference between the "natural" and the "naturalistic" attitude. In conceptualizing the patient's illness in terms of objective, scientific constructs, the physician remains within the "naturalistic" attitude. In the "naturalistic" attitude the physician, in effect, reifies the illness and conceives of it as an objective entity – a disease state. That is, the purpose in the "naturalistic" attitude is to grasp the patient's illness as a pathological "fact." As Richard Baron (1981, p. 7) has noted, the prevailing commitment to accurate diagnosis of disease – which is the hallmark of the modern physician – turns on the notion that there is a pure disease state which is, ideally, distinct from the patient.[36] Thus, the patient is seen as a kind of "translucent screen" on which the disease is projected. In consequence, in the "naturalistic" attitude, the patient's subjective experiencing of illness is ignored in favor of an objective, quantitative account of a disease state. As Michel Foucault comments:

In order to know the truth of the pathological fact, the doctor must abstract the patient.... Paradoxically, in relation to that which he is suffering from, the patient is only an external fact; the medical reading must take him into account only to place him in parentheses. Of course, the doctor must know "the internal structure of our bodies"; but only in order to subtract it, and to free to the doctor's gaze "the nature and combination of symptoms, crises and other circumstances that accompany diseases." It is not the pathological that functions, in relation to life, as a *counternature*, but the patient in relation to the disease itself (Foucault, 1975, p. 8).

Thus, disease, as it is conceived within the "naturalistic" attitude, represents a scientific abstraction from the immediate lived experience of the patient.

## Temporality

With regard to the lived experience of illness, it is important to recall the distinction between lived (or subjective) time and objective time. The patient experiences his or her illness in its immediacy in terms of the ongoing flow of "lived" time. If one is in pain, for example, each flicker of pain does not represent a discrete, atomic instant along a time-line, but rather a continuum of discomfort in which past and future pains coalesce into a stagnating present. As Calvin Schrag notes with regard to this temporality:

> The moments of pain ... do not follow the regular and ordered sequence of seconds and minutes that are marked off by the swing of a pendulum or the ticking of a clock. Clock time is isotropic. The values of its units are uniform. The time of one's being in pain is anisotropic. Its values vary with the intensity of the pain, the accompanying emotional weight, and the press of concerns at hand (Schrag, 1982, p. 122).

Illness as it is "lived through" is experienced as an ever-present, enduring consciousness of disorder which resists measurement in terms of objective time. In the preoccupation with the here and now, the person who is ill pays little attention to clock time. Minutes may seem like hours, hours like days.[37] The person who is ill is like Schutz's beholder of the musical piece. Just as the beholder of the musical piece has little awareness of clock time while listening to the music, and hence he or she may be surprised later to learn that one movement takes exactly as much clock time as another, so the person who is ill has little awareness of clock time in actually living through the discomfort.

The physician, on the other hand, uses the objective time scale to measure the physical events and biological processes which define the patient's illness as a disease state (and to plan therapeutic interventions). Consequently, physician and patient are constituting the temporality of illness and the disease state according to two different and incommensurable time dimensions.[38]

*Relevance*

Schutz has emphasized that what one attends to depends upon the project in which one is engaged and the system of relevances that are a function of one's life plan. In particular, Schutz (1962d, pp. 226–229) notes that the world of everyday life is governed by the pragmatic motive. As such it is organized into strata of major or minor relevance. The individual selects as being of primary importance those objects which actually are or will become in the future possible ends or means for the realization of projects, or which are or will become dangerous or enjoyable or otherwise relevant. Furthermore, Schutz argues that the overall system of relevances within the practical sphere of everyday life is governed by what he has termed the "fundamental anxiety" – "the basic experience of each of us: I know that I shall die and I fear to die."[39]

While engaging in the scientific project, the scientist adopts a system of relevances which are governed by "stating the problem at hand" (Schutz, 1962b, p. 37). In "stating the problem at hand," the scientist defines what is considered relevant and guides the process of inquiry. Consequently, the system of relevances changes. What is relevant to the scientist in engaging in scientific work may be quite irrelevant in his or her daily life, and vice versa.

In attending to the experience of illness the physician does so in light of scientific training and the goals of the profession. In so doing, he or she focuses on the disease process itself. Consequently, the clinical data are of highest relevance. However, the patient is less concerned with the objective clinical data. What is most relevant to the patient is the effect the illness will have upon daily life.

Leo Tolstoy has captured this shift in orientation in *The Death of Ivan Ilych*:

To Ivan Ilych only one question was important: Was his case serious or not? But the doctor ignored that inappropriate question. From his point of view it was not the one under consideration, the real question was to decide between a floating kidney, chronic catarrh, or appendicitis.... All the way home (Ilych) was going over what the doctor had said, trying to translate those complicated, obscure, scientific phrases into plain language and find in them an answer to the question: "Is my condition bad? Is it very bad? Or is there as yet nothing wrong?" (Tolstoy, 1978, pp. 520–521).

Walker Percy (1954, pp. 125–126) has elaborated on this distinction by suggesting that one may differentiate between "knowledge sub specie aeternitatis" (knowledge that can be arrived at anywhere by anyone at any time, e.g., the boiling point of water) and "news" that expresses a "contingent and nonrecurring event or state of affairs which ... is peculiarly relevant to the concrete predicament of the hearer of the news." The significance of a statement for an individual will depend upon his or her situation. "To say this," says Percy, "is to say nothing about the truth of sentences. Assuming that they are all true, they will have a qualitatively different significance for the reader according to his own placement in the world" (Percy, 1954, p. 128). For example, the castaway on a desert island and the individual in the midst of civilization will react differently to the statement "there is water over the next hill." While the statement may be of momentous import to the one, it may be of little interest to the other.

The scientist has abstracted from his or her own existential situation in order to achieve objectivity. What is significant to the scientist as a piece of "knowledge" may be significant to another as a piece of "news"; that is, the information may have a peculiar relevance for the other's concrete predicament in the world. Such is the case in the patient-physician encounter. The clinical data represent "news" to the patient and "knowledge" to the physician. Each, therefore, reacts to the information in a distinctly different manner.

In this regard Cassell (1979, p. 203) notes that the patient is both experiencer and "assigner of understandings." The meaning of illness to a particular patient will depend upon "the collectivity of his meanings" – a collectivity that is necessarily a function of his or her unique biographical situation. Thus, an experience of pain might be interpreted by one patient as a possible heart attack and by another as merely indigestion. The significance of the pain to the particular patient will depend upon his or her life history and the personal meanings constituted within that life history. Likewise the significance of the clinical data to the particular patient will depend upon his or her unique biographical situation. A clinical diagnosis may be regarded as "terrible" by one patient, and as merely "inconvenient" by another. Each reacts to the "news" of the diagnosis according to its peculiar relevance to his or her concrete situation within the world. In this connection it will be recalled that the present is constituted in light of past experiences and future anticipations which relate to the individual's unique life plan.

As Cassell (1979, pp. 204–205) points out, the physician is also an "assigner of understandings" in that the physician takes the patient's subjective report of illness and reinterprets it in terms of his or her understanding of disease processes. That is, in attending to the patient's account of the lived experience of illness, the physician interprets this account in terms of his or her knowledge of physiology, anatomy, and so forth (i.e., the patient's lived experience is placed within the "naturalistic attitude") in order to determine therapeutic interventions. This assignment of meaning on the part of the physician will, however, be quite different from the patient's assignment of meaning.

The physician defines the "problem at hand" in light of certain goals of medicine: diagnosis, treatment and prognosis. These goals appear to be shared with the patient. However, as Baron (1985, p. 609) notes, the patient defines the "problem at hand" in terms of different goals. What the patient seeks is explanation, cure and prediction. This is not the same thing. The patient's goals relate to the qualitative immediacy of illness. They represent an attempt to integrate the experience into daily life. In seeking explanation, the patient seeks a validation of the illness experience, a means reasonably to account for the feeling that something is wrong ("You have a pain because you have gallstones.") If no explanation is forthcoming ("Your tests are negative. I can't find anything wrong with you.") the patient is at a loss as to how to make sense of the illness. In seeking a cure, the patient anticipates a perfect restoration of health, a return to the way things were before he or she became ill. In asking for a prognosis, the patient expects a prediction of what is going to happen to him/her personally.

Baron suggests, however, that the physician's goals of diagnosis, treatment, and prognosis usually represent "derivative or secondary goals":

Diagnosis for us is categorization (for example, acute promyelocytic leukemia or acute myelomonocytic leukemia), not explanation. Treatment virtually never results in cure, if only because treatment itself usually has an effect on people's lives, altering "the way things were before." Prognosis is always statistical and in that sense rarely tells a particular person what will happen to him or her (Baron, 1985, p. 609).

Since the "problem at hand" is defined differently by patient and physician, according to goals that relate to their separate worlds, they do not share a system of relevances with respect to these goals. It is clear, for example, that the system of relevances of the patient is governed in a very explicit way by the "fundamental anxiety." The experience of illness is such that one comes face-to-face with one's own inherent vulnerability. One reacts to the "news" of illness accordingly. The physician, on the other hand, is attending to the patient's experience of illness according to the system of relevances which are a function of scientific training and the goals of the profession. In this respect the physician prescinds from the fears and anxieties of the patient and focuses upon the clinical data as "knowledge" rather than "news," in order to determine what actions or medical interventions may be possible. Consequently, most physicians tend to act upon a narrower range of goals than the patient and to address a sub-set of the patient's concerns.[40]

### *"Communicative Common Environment"*

In considering the manner in which patient and physician constitute the meaning of illness differently, it is thus important to note that the experience of illness is such that the factors which Schutz has identified as integral to a common understanding no longer provide the means to constitute a "common world." The patient and physician find it difficult to communicate about the experience of illness on the basis of a shared set of assumptions.

As Schutz notes, to be successful the communicative process must involve a set of common abstractions or standardizations. Typification is a form of abstraction which provides the basis for standardization within common-sense thinking. Indeed, the individual ordinarily interprets daily life in light of typifications which make up the stock of knowledge-at-hand and which render experience predictable and controllable. By means of such naive typifications the familiar world assumes a quality of taken-for-grantedness such that one expects things will continue more or less as they have proven to be in the past. This taken-for-grantedness permeates the fabric of daily life and it is on the basis of a typified stock of knowledge-at-hand that a shared world of experience is possible. In particular, shared typifications provide the ground for successful communication.

In the patient-physician encounter it is often the case that doctor and patient do not communicate on the basis of a shared set of typifications. In the first place the lived experience of illness is such that, whereas the doctor interprets the patient's illness as a typified instance of a particular disease state, the patient encounters the disorder as a unique personal event. As has been noted earlier in this work, this is an explicitly different focus. Typification involves grasping an object as an example of a certain type (e.g., as a tree, an automobile, a mountain, and so forth). Thus, to grasp a particular instance of illness as a typification is to consider it not in its uniqueness but rather as it exemplifies something other than itself (e.g., as a typical case of diabetes, chicken pox, measles). In *living through* illness the patient does not experience bodily disorder as simply exemplifying a typified instance of disease. Rather, one experiences illness in the unique manner in which it impacts upon one's particular life situation (Cassell, 1979, pp. 203–205). The temporality of lived experience is important in this respect. To grasp something as a typification is to consider it apart from its ongoing constitution in lived time.[41]

Furthermore, the impact of illness on the life of the patient may be such that he or she finds it difficult readily to incorporate the experience into the mundanity of everyday existence. This is perhaps especially the case if the illness comes on unexpectedly or if it appears to be serious in nature. In the experience of illness the taken-for-granted quality of daily life is called into question. What is primarily threatened is the integrity of the self (one's own self), and this most fundamental loss of wholeness (this ontological threat) cannot readily be interpreted in terms of naive typifications. The most deeply held assumption of daily life is the assumption that I, personally, will continue to be alive and it is in light of this assumption that one engages in daily activities.[42] The onset of illness, however, brings one concretely face-to-face with personal vulnerability. The loss of control that is intrinsic to the experience of illness is accompanied by an acute awareness of the unpredictability of the familiar world. It can no longer be assumed that things will continue much the same as they have in the past. Thus, the person who is ill finds prior assumptions about the familiar world, the stock of knowledge-at-hand, to be strangely inadequate for interpreting the existential crisis. The patient is unable readily to fit illness into the typified schema used to organize and interpret experience.[43]

Consider, for example, the distinction between death as a typified event and death as a personal, concrete awareness that I, myself, will no longer continue to be alive, as portrayed in the following quote from Heidegger:

In the publicness with which we are with one another in our everyday manner, death is "known" as a mishap which is constantly occurring – as a "case of death." Someone or other "dies," be he neighbor or stranger.... People who are no acquaintances of ours are "dying" daily and hourly. "Death" is encountered as a well-known event occurring within-the-world. As such it remains in the inconspicuousness characteristic of what is encountered in an everyday fashion.... The analysis of the phrase "one dies" reveals unambiguously the kind of Being which belongs to everyday Being-towards-Death. In such a way of talking, death is understood as an indefinite something which, above all, must duly arrive from somewhere or other, but which is proximally not yet present-at-hand for oneself and is therefore no threat. The expression "one dies" spreads abroad the opinion that what gets reached, as it were, by death is the "they." In Dasein's public way of interpreting it, it is said that "one dies," because everyone else and oneself can talk himself into saying that "in no case is it I myself" (Heidegger, 1962, pp. 296–297).

Since the lived experience is unique in its existential impact, the ill person often finds it difficult to communicate the experience to others on the basis of a shared set of typifications.

The physician, on the other hand, *is* able to interpret the illness of the patient in terms of his or her stock of knowledge-at-hand and may be unaware that the patient does not conceive of the illness as a typification. Moreover, the typifications which the physician uses to characterize the patient's illness are significantly different from those naive typifications which characterize daily life. The doctor, as does any specialist, acquires through training a whole new set of types.[44] Such scientific typifications are characteristic of what Schutz (1973, pp. 314–315) has called an "autonomous province of knowledge." Autonomous provinces of knowledge are to be distinguished from general knowledge (i.e., knowledge routinely transmitted to everyone and, in principle, accessible to all) in that the acquisition of such specialized knowledge requires complicated learning sequences which precede its acquisition. Consequently, such specialized knowledge is only readily available to "experts."[45] Thus, not only is it the case that the physician sees the patient's illness as a typification, but further that he or she conceives of it in terms of scientific typifications – the latter being significantly different from the types used by the patient in everyday life.

While it is the case that one always lives through one's particular illness in its uniqueness, at the reflective level one may typify certain aspects of the illness experience. One might, for example, note that a feeling of feverishness, a sore throat, and a general ache in one's muscles is "typical" of the onset of the 'flu. Indeed, as a multiple sclerosis patient, I can conceive of certain of my permanent and ongoing disabilities (such as loss of equilibrium, gait disturbance, and so forth) as "typical" of my disorder. Nevertheless, it should be noted that these typifications still relate in large part to *my* unique experience of the illness (rather than representing for me typified instances of an abstract, objective disease process). The loss of equilibrium manifests itself as "typical" of my experience of M.S. in the sense that I must routinely hold on to furniture or touch walls as I make my way around the house (and, more particularly, I must steady myself on the wooden bookcase in the hallway and catch hold of a certain chair as I move across the den). My gait disturbance is "typical," not only in that many M.S. patients experience such disturbance, but in the sense that my illness incorporates a "typical" way of being for me – a way of being in which walking is effortful, uncoordinated, and accomplished only with the aid of crutches or a walker.

In considering the communicative process between doctor and patient, then, it is important to recognize the role of typification in achieving successful communication. To what extent do patient and physician share a set of common abstractions or standardizations? Obviously, to some extent doctor and patient share a set of typifications (i.e., prescientific typifications) on the basis of which the patient attempts to describe the experience of bodily disorder and the physician begins the diagnostic process. That is, the patient attempts to describe the atypicality of his or her experience in terms of its deviation from typical ways of being and the physician attempts to grasp this atypicality in a "naive" way prior to interpreting it in light of scientific knowledge. As Michael Schwartz and Osborne Wiggins (1985, p. 354) have noted, the physician achieves an understanding of illness on the basis of prescientific typifications (e.g., the scientific notion of emphysema presupposes an ordinary understanding of breathlessness). In adopting the scientific attitude, however, the physician moves to the level of scientific typifications (typifications which are not normally shared by the patient) and conceptualizes the patient's illness as representing a more or less typified instance of an objective, disease process. At this level patient and physician no longer communicate on the basis of a shared set of typifications. Furthermore, though attempting to describe it in typified terms, the patient always experiences illness in its uniqueness.

In communicating about illness, however, patient and physician assume that they are discussing a shared reality, a common object. This assumption is made on the basis of the two idealizations of the "general thesis of reciprocal       perspectives."       Through       the       idealization       of       the

"interchangeability of the standpoints" the individual takes for granted –
and assumes that his fellow man does the same – that if they were to
change places then each would see essentially what the other now sees.
Through the idealization of the "congruency of the system of relevances"
the individual takes for granted that the difference in perspectives
originating in the unique biographical situation of himself and his fellow
man is irrelevant for the purpose at hand, and that both he and his fellow
have selected and interpreted common objects in an identical manner or,
at least, an "empirically identical" manner sufficient for all practical
purposes.

Thus, the patient and physician both assume that, in communicating
about the illness, they are doing so on the basis of a shared understanding,
that they are interpreting illness in an "empirically identical" manner. The
patient takes for granted that the physician recognizes the illness as
primarily and essentially a threat to his or her personal being. The
physician assumes that the patient understands the disease (albeit
incompletely) in terms of the "objective" clinical data. Thus, the constructs of common sense thinking, rather than enabling the patient and
physician to share a common "reality" tend to deepen the chasm between
their separate worlds.

This failure of the "interchangeability of standpoints" is not, however,
simply a matter of disparate interpretations of a common object.[46] In a
more fundamental respect the lived experience of illness is such that it
cannot represent a common object. Illness is, first and foremost, a
subjective experience. As such, it is an inner – rather than an outer –
event which, in large part, cannot be shared with another. It is, for
example, by no means evident that if I changed places with you, then I
would have substantially the same experience as you are having when it
comes to such inner experiences as pain. For where is pain located? It is
not an object like a cup on the table which you and I may both perceive
and which (if we were to change places) we would presumably perceive
in essentially the same manner.[47] There is, thus, an unshareability
characteristic about illness which derives from its being an inner, rather
than an outer, event.[48] Indeed, as Elaine Scarry has noted, there seems to
be no language adequate for communicating such inner events as pain to
another who does not share that inner event:

[W]hen one speaks about "one's own physical pain" and about "another person's
physical pain," one might almost appear to be speaking about two wholly distinct

orders of events. For the person whose pain it is, it is "effortlessly" grasped (that is, even with the most heroic effort it cannot *not* be grasped); while for the person outside the sufferer's body, what is "effortless" is *not* grasping it (it is easy to remain wholly unaware of its existence; even with effort, one may remain in doubt about its existence or retain the astonishing freedom of denying its existence; and, finally, if with the best effort of sustained attention one successfully apprehends it, the aversiveness of the "it" one apprehends will only be a shadowy fraction of the actual "it") (Scarry, 1985, p. 4).)

"Whether or not pain is the most difficult part of cancer to live with," says Arthur Frank (1991, p. 29), "it is probably the hardest to describe."

We have plenty of words to describe specific pains: sharp, throbbing, piercing, burning, even dull. But these words do not describe the experience of pain. We lack terms to express what it means to live "in" such pain. Unable to express pain, we come to believe there is nothing to say. Silenced, we become isolated in pain, and the isolation increases the pain. Like the sick feeling that comes with the recognition of yourself as ill, there is a pain attached to being in pain (Frank, 1991, pp. 29–30.)

This unshareability aspect of illness results not only in the failure of the "interchangeability of standpoints" but also in an incongruence in the appresentational schema of physician and patient. In some sense it appears that in the communicative process between doctor and patient language appresents two distinct entities. For the patient language is intended to appresent (albeit inadequately) the inner event of illness; for the doctor language appresents the disease lurking behind the patient's subjective experience. That is, the object for which the language is presumed to be an appresentational reference is not a common object.

An important factor that contributes to the unshareability aspect of illness is the incommensurability of inner and outer time (or lived and objective time). The patient must describe the illness in terms of outer time (since this is the common language for time). Yet one experiences one's illness in its immediacy in terms of inner time. The reference to outer time represents an interpretive scheme imposed upon the lived experience. The necessity for using the objective time scale as a means for communicating the lived experience of inner time creates difficulties for the person attempting to communicate the experience of illness. It is often hard to gauge the duration of alien body sensations when one is living through such sensations.

As Schutz has noted, successful communication also presupposes a certain taken-for-granted congruence in the interpretational schemes of the communicators; that is, the communicator assumes that the interpreter will interpret his or her communicative sign in substantially the way that he/she interprets it. In the doctor-patient relationship this assumption is problematic since the communicator (the patient) intends the communicative sign to relate to his or her subjective experience of illness and the interpreter (the physician) is interpreting the communicative sign as relating to disease. As Schutz indicates, full identity of the interpretational schemes is impossible in principle (since such interpretational schemes are determined by the unique biographical situation of the communicators). Any successful communication is, thus, dependent upon the communicators sharing a substantially similar system of relevances. As has been noted earlier, however, doctor and patient do not share a substantially similar system of relevances with regard to the patient's illness.

## Summary

In sum, then, the foregoing analysis reveals that there is a tendency systematically to distort meaning in the doctor-patient relationship. In everyday life a common world is ordinarily constituted through the establishment of a "communicative common environment." Such a "communicative common environment" is possible because certain constructs of common sense thinking and such factors as shared typifications, congruent interpretational schemes, and substantially similar systems of relevances to a large extent overcome differences in individual perspectives. The failure of the general thesis of reciprocal perspectives (which is grounded in the unshareability aspect of illness) and the incongruence between the typificational, interpretational and appresentational schema of doctor and patient present particular difficulties for the establishment of a shared world of meaning.

### 4. IMPLICATIONS FOR MEDICAL PRACTICE

The phenomenological analysis of the constitution of meaning provides some practical insights for those engaged in medical practice. In the first place, such an analysis reveals that the difference in perspectives between physician and patient is much more serious than is generally recognized.

It becomes clear that such a difference in perspectives is not simply a matter of varying levels of knowledge – as is often assumed to be the case – but rather such a difference is grounded in the distinction between lived experience and scientific conceptualization. The patient necessarily experiences illness in its immediacy. To the extent that the physician conceives of the illness purely as a scientific construct (i.e., as a disease state), so he or she moves away from the patient's immediate experience and there opens a decisive gap in understanding between them.

In revealing the primacy of lived experience over and above any subsequent theoretical scientific account of such experience, the phenomenological account discloses the validity of the patient's subjective experience of illness. Critics of modern medicine argue that such subjective experience on the part of the patient is often discounted as unreliable and treated as "soft data" to be essentially ignored in favor of the "hard," objective, quantitative data of laboratory tests, x-rays, and so forth (Schwartz and Wiggins, 1985; Engel, 1977b; Baron, 1985; Donnelly, 1986). The foregoing analysis shows that the patient's experience must be taken into account not simply as a subjective accounting of an abstract "objective" reality but rather that the patient's experience must be taken into account because *lived experience represents the reality of the patient's illness*.

The fundamental insight that in the constitution of meaning there is an essential correlation between the experiencer and that which is experienced underscores the necessity of considering the manner in which each patient constitutes the meaning of his or her personal experience of illness. No two patients will assign exactly the same meaning to their disorder. Thus, it becomes of vital importance to consider the "horizon" of the patient's world in terms not only of a unique biographical situation but also in terms of the wider social meanings which are a function of particular ethnic and cultural backgrounds. For example, as physicians such as Cassell (1979) and Arthur Kleinman (1988) have shown, the meaning of a particular illness to a particular patient will depend upon the "collectivity of his meanings" – a collectivity which is in part determined by social meanings. Acknowledging the cultural background of the patient is imperative since symptoms of illness have a different significance according to particular ethnic and cultural backgrounds, as well as according to the personal meanings embedded in a particular life narrative (Kleinman, 1988, pp. 21–24, 100–120).

The physician, as an experiencer, is of course equally an assigner of meaning to the illness – an assignment which derives from *his or her* biographical situation and which necessarily differs from that of the patient. Studies suggest that an explicit recognition that this is the case can prove invaluable in enabling the physician to begin the task of constructing a shared world of meaning with the patient. The doctor who monitors personal reactions and feelings towards the patient and the patient's illness, is better able to recognize and set aside any preconceived notions which may impede the ability to explore the meanings inherent in the patient's world (Elder and Samuel, 1987). In this respect it is important to note that, in face-to-face contact with patients, physicians may come to experience illnesses as "frustrating," "boring," "interesting," "a limitation on one's capacities," "a challenge to one's expertise," and so forth. In other words, the physician's lived experience of the patient's illness is significantly different from the patient's lived experience of the patient's illness.

The importance of understanding the patient's lived experience should not be underestimated. Therapy is less likely to be successful if the physician fails to take into account what the illness means to the patient (Kleinman, 1988, pp. 239–241). Indeed, as Cassell (1982, pp. 639–645) has shown, one cannot begin to address the patient's suffering unless attention is paid to such meaning. On the other hand, case studies demonstrate that physicians who explicitly focus on the meaning of the patient's experience find they are better able to care for their patients. For example, Hoyle Leigh and Morton Reiser (1980, pp. 243 ff.) show how the understanding of a patient's experiencing allowed his physicians to treat him more effectively in the intensive care unit. Cassell (1985a, pp. 157 ff.) discusses a patient with intractable pain in the throat whose father died of cancer of the esophagus. Understanding the meaning of the pain to the patient enabled Cassell to alleviate the physical symptoms of the patient's illness. In exploring the meanings inherent in accounts of particular illness experiences of chronically ill patients, Kleinman (1988) proposes a practical methodology for treating those with chronic illnesses. In his clinical case histories, Oliver Sacks (1985c) demonstrates that the "human vision" of the physician (as opposed to his "medical vision") can provide invaluable insights into the patient's particular situation. Such insights are not readily apparent from a review of the clinical data alone. The physician's "medical gaze" is directed at the clinical picture; his "human vision" is focused on the person who is ill.

Certain matters which are of perennial concern to physicians, such as the apparent non-compliance of a large number of patients, may be more readily understood once one recognizes that many times physicians and patients do not share a system of relevances. It is not necessarily the case that what the physician deems good for the patient is the same as what the patient considers best for himself. Treatment decisions, therapeutic goals, value choices, estimations of what is ultimately in one's own best interest, are all affected by the system of relevances which are a function of one's particular life plan.[49] Consequently, it is a matter of some importance that physician and patient make it clear to each other just what each considers to be of primary importance in the therapeutic endeavor so that they may negotiate and collaborate on a system of care.

As has been noted, one of the difficulties inherent in patient-physician communication arises from the fact that illness in its immediacy is a subjective (inner) experience. Consequently, it is not easy for patients to communicate this experience to others. In particular, it is often hard for patients to give an account of their illness according to the units of the objective time scale. This difficulty leads to the distrust of the patient as a reliable narrator and the patient's experience is bypassed in favor of what is taken to be a more "objective" rendering of the disease state. The phenomenological analysis shows, however, that to bypass the patient's voice is to bypass the illness itself. The recognition that the apparent non-specificity of the patient's account is simply a reflection of the un-shareability aspect of illness motivates a different approach to the patient's narrative. The narrative is seen to be central to an understanding of the patient's illness. Indeed to understand the illness-as-lived, Baron (1981, p. 19) suggests, the physician must "go beyond questions such as 'When did it begin? Do you have black, tarry stools? Does it get worse when you walk?' and develop such questions as 'What is it like?' or 'How is it for you?'"

Indeed, diagnostic questionnaires such as the McGill Pain Questionnaire take seriously the linguistic difficulty of communicating the felt experience of pain and reject conventional medical vocabulary ("moderate pain," "severe pain") in favor of groups of adjectives such as "flickering," "quivering," "pulsing," "throbbing," and "beating" (adjectives often spoken by patients) to aid patients in more readily generating descriptions of their experience.

In this respect it is important to note the difference between interrogation and dialogue. Questions which admit of only "yes" or "no" answers do not allow the respondent to provide a description of his or her experience. Simply to answer "yes" or "no" to the question "Are you experiencing numbness?" omits the lived experience altogether. This is particularly the case if the interrogation is premised on a purely "biological" perspective and questions are intended to elicit "objective" data. If the physician is to learn something about the patient's experience, he or she must initiate a dialogue with the patient – a dialogue that allows the patient to provide a first person narrative of the illness.

The unshared aspects of illness may be minimized in other ways. Literature (plays, novels, short stories) and personal accounts written by patients can provide information which is otherwise not readily available to the physician. Indeed, Baron (1985, p. 609) argues that many works of literature may be read as "medical treatises that give physicians information absolutely essential to the practice of medicine." Literary descriptions of illness (whether fictional or autobiographical) provide insight into the existential predicament of illness – what it is like to be sick. Physicians who have themselves been ill (or who have experienced illness in their families or in those close to them) find they have a greater understanding of their patients' situation (Rosenbaum, 1988; Mandell and Spiro, 1987; Kleinman, 1988, pp. 211–213). Literature can provide similar insights for the uninitiated and may aid in the development of a shared world of meaning between physician and patient.

It is only by understanding what it is that keeps them apart, that physician and patient may take concrete steps to build bridges between their separate worlds. The analysis of "communicative common environment" suggests that attention be given to recognizing such differences as disparate systems of relevances, different habits of mind, distinct typificational and interpretational schemata, the unshareability character of illness, and so forth. Such differences need to be made explicit if they are to be confronted and resolved. As has been noted, a fundamental difference which exists is that between the lived experience of illness and the conceptualization of illness as a disease state. This difference will be further explored in Chapter Two.

# ILLNESS

### 1. LEVELS OF CONSTITUTION OF MEANING

The phenomenological analysis of the "worlds" of physician and patient reveals a fundamental distinction between the lived experience of illness and its conceptualization as a disease state. In particular, a distinction has been noted between meaning which is grounded in lived experience and meaning which represents an "abstraction" from lived experience.

In an attempt to elucidate the manner in which illness is apprehended differently by physician and patient, it is helpful to consider Jean-Paul Sartre's (1956a, pp. 436–445, 463–470) analysis of pain and illness in which he identifies four distinct levels of constitution of meaning: pre-reflective sensory experiencing, "suffered illness," "disease," and the "disease state." The first three represent the manner in which the patient apprehends illness; the "disease state" represents the physician's conceptualization of illness.[50]

Sartre (1956a, pp. 436–438) argues that the fundamental level of constitution of illness is that of pre-reflective sensory experiencing. At this level the immediate, pre-reflective experiencing is a manifestation of the way consciousness "exists" the body. A pain in the eyes, for example, is not immediately experienced as an object "pain" which is located *in* the eyes. Rather, pain *is* the eyes at this particular moment. One experiences the eyes-as-pain, vision-as-pain, the peculiar contingency of, say, this particular act of reading which manifests itself in terms of the blurring of the words, the inability to concentrate on this particular passage in the book, and so forth.

In contrast, Sartre (1956a, pp. 440–441) says, if I reflect on my pain and attempt to apprehend *it*, the pain ceases to be lived-pain and becomes object-pain. In the reflective act, the pure quality (consciousness) of pain is transcended by a psychic object, "suffered illness." As lived unreflectively (or pre-reflectively) the pain *is* the body. When reflected upon, pain becomes a psychic object (illness) outside one's immediate subjectivity and thus becomes identified as, say, pain "in the stomach." For the reflective consciousness, then, illness is *distinct from* the body and has its own form. At this point "each concrete pain is like a note in a melody: it is at once the whole melody and a 'moment' in the melody" (Sartre, 1956a, p. 442). With each pain one apprehends the illness and yet "it transcends them all, for it is the synthetic totality of all the pains, the theme which is developed by them and through them" (Sartre, 1956a, p. 442).

At yet another level of reflection one apprehends one's illness as "disease." At this level illness represents an objective disease, such as ulcer of the stomach, which is known by means of bits of knowledge acquired from others (i.e., such knowledge as the principles of physiology and pathology described by others) (Sartre, 1956a, p. 466). In its immediacy illness is not *experienced* as "disease." It is, rather, experienced as the body painfully-lived. In this regard Sartre is concerned to show that since I *am* my body, the lived body is an inapprehensible given which is never grasped as such. In the normal course of events I do not experience my body as a neurophysiological organism (i.e., as a skeleton, brain, nerve endings, and so forth). It is only if I conceive of my body as an object (in Sartre's terms, as a "being-for-others") that I apprehend it as a malfunctioning physiological organism. "Disease" represents such objectification. The immediate experiencing of my stomach painfully-lived is grasped by me not only as pain "in the stomach" but, further, as "gastralgia." Furthermore, this conception of "disease" incorporates the knowledge of a certain objective nature possessed by the stomach:

I know that it has the shape of a bagpipe, that it is a sack, that it produces juices and enzymes, that it is enclosed by a muscular tunica with smooth fibres, etc. I can also know – because a physician has told me – that the stomach has an ulcer, and again I can more or less clearly picture the ulcer to myself. I can imagine it as a redness, a slight internal putrescence; I can conceive of it by analogy with abscesses, fever blisters, pus, canker sores, etc. All this on principle stems from bits of knowledge which I have acquired from Others or from such knowledge as Others have of me. In any case all this can constitute my Illness, not as I enjoy possession of it, but as it escapes me (Sartre, 1956a, p. 466).

The level of the "disease state" identified by Sartre represents the physician's conceptualization of the patient's illness. Illness is identified with a pathoanatomical or pathophysiological fact. Sartre (1956a, p. 466) notes that illness is thereby wholly conceived as "a question of bacteria or of lesions in tissue."

## 2. THE PATIENT'S APPREHENSION OF ILLNESS

Sartre suggests that the fundamental level of constitution of the meaning of illness is that of pre-reflective sensory experiencing and it is indeed usually the case that the patient first becomes aware that all is not well in the felt experience of some alien bodily sensation (such as pain, itch or chill) or in the sensed experience of a change in function (such as the unusual weakness of a limb, the abnormal stiffness of joints or the unaccustomed loss of coordination). Additionally, one might become aware of an alteration in the normal appearance of one's body – some disfigurement such as a rash or a lump – which would lead to the apprehension of illness at the reflective level.[51]

It is, of course, the case that a sensory experience such as pain is not *always* apprehended at the reflective level as "suffered illness" – for example, if I drop a brick on my toe or if I have a headache following an evening of heavy drinking, I may experience pain but do not conceive it to be illness. Such would also be the case if I experienced a weakness in my legs and an abnormal redness in my face following a long and vigorous game of tennis, or an ache in my back after digging in the garden. Nevertheless, more often than not, the apprehension of "suffered illness" at the reflective level is the result of immediate sensory experiencing at the pre-reflective level.

In this regard Cassell (1985a, p. 25) notes that symptoms of illness are the patient's reports of what is experienced as an alien body sensation. He points out that the key point is that the sensation is experienced as *alien* or *unusual*. Not all abnormalities are symptoms in that, if the person has become acclimatized to the abnormality, then it is no longer regarded as an alien body sensation – and hence as a symptom. As an example Cassell notes that heavy smokers may deny that they have a cough, even though one may hear them coughing. "Cigarette cough" has become part of them. It is a way of life and, since it is not experienced at the pre-reflective level as an alien sensation, it is not apprehended as "suffered illness" at the reflective level. H. Tristram Engelhardt (1976, p. 260) notes that, in order for pre-reflective sensory experience to be apprehended as illness at the reflective level, it must be perceived to be dysfunctional or to involve pain which is not a part of a function deemed proper to the human organism (for example, consider the pain of teething as compared to the pain of a migraine headache).

If the immediate experience of bodily disruption is sufficiently unusual, prolonged, uncomfortable, and so forth, then it must be explicitly attended to by the patient and reflected upon. Consequently, at this point, the experience becomes one that must be given meaning (Cassell, 1985a, p. 26). In focusing on the unusual sensory experience, the patient's attention shifts to the body and the bodily disruption is itself made thematic. At the pre-reflective level the body is not explicitly given to consciousness. Rather one is engaged in the world, preoccupied with one's projects. The body is "surpassed" in carrying out one's projects in the world. For example, if I am reading a book, my attention is wholly directed to the meaning of the text. I am not explicitly aware of the functioning of my eyes in the act of reading. It is rather the meaning of the text itself that is thematic to my consciousness. However, if I have a headache and reading becomes difficult, then my attention is diverted from the meaning of the text. I focus instead on the act of reading and I attempt to identify the source of the difficulty. I become aware that the source of the difficulty is pain and, furthermore, that this pain is located "in the eyes." At this reflective level Sartre argues that the immediate pre-reflective sensory experiencing is thus apprehended as "suffered illness" – a psychic object, an "it" which is somehow distinct from the body. That is, rather than being simply experienced as the eyes painfully-lived, pain becomes a separate entity which is located "in the eyes." The expression "painfully-lived" here should not be taken to imply that illness at the pre-reflective level always involves the experience of pain, per se. Obviously, the pre-reflective experience of "illness" may not include pain although it will involve some felt bodily disruption or some perceived disfigurement which causes the body to become thematic in a distinct way.

Cassell notes that patients often refer to this psychic object (the "it") when attempting to communicate their lived experience of illness:

And then it seemed to me that the gripping shifted.... It'd go on for a minute or two ... then it shifted to a lower part.... And then in the morning it persisted ... but it seems to be centered around here, in the middle of the stomach (Cassell, 1985a, p. 14).

This conception of sensory experiencing as a distinct entity is well reflected in Ivan Ilych's description of the pain of his fatal illness:

But suddenly in the midst of those proceedings the pain in his side ... would begin its gnawing work. Ivan Ilych would turn his attention to it and try to drive the thoughts of it away but without success. *It* would come and stand before him and look at him, and he would be petrified and the light would die out of his eyes, and he would again begin asking himself whether *It* alone was true. And his colleagues and subordinates would see with surprise and distress that he, the brilliant and subtle judge, was becoming confused and making mistakes.... And what was worst of all was that *It* drew his attention to itself not in order to take some action but only that he should look at *It*, look it straight in the face: look at it without doing anything, suffer inexpressibly (Tolstoy, 1966, p. 528).

This description well illustrates the shift of attention which occurs in the presence of unusual sensory experience. The lived experience of the body itself becomes the focus of attention.[52] Pain or other bodily dysfunction disrupts one's ongoing engagement in the world. The body can no longer be taken for granted and ignored. Rather, the bodily disruption must be attended to and interpreted.

In addition, at this reflective level, there is an intuitive awareness on the part of the patient that the symptoms are part of a larger whole. That is, the various isolated bodily disturbances point to, or signify, a more complex entity of which they are simply one phase or facet (i.e., they are not experienced as discrete sensations bearing no relation to a larger unity). With each pain or disruption one apprehends the illness, and yet it is a synthetic totality which transcends them all. Ilych, for example, was intuitively aware that the pain in his side (the "it") was but one facet of an even more complex and dreadful reality – a disease which was killing him.

Sartre argues that at this point illness is still an immediate lived experience. "Suffered illness" manifests itself *as* the collection of alien body sensations (for Ilych the totality of pains) which disrupt sensory experiencing at the pre-reflective level (i.e., at this level it is not apprehended as a *particular* illness – that comes at the level of "disease.")

At the level of "disease" the patient experiences his or her body as an object (i.e., as a neurophysiological organism which possesses a certain objective nature). Thus, in Sartre's terms, "disease" represents a "being-for-others" in that it is known to the sick person by means of concepts derived from others. Furthermore, the disruption in everyday experiencing (the "suffered illness") is now apprehended by the patient as being a "disease" – an abstract entity residing in but in some way distinct from the body.

In this regard Engelhardt (1982, p. 146) has pointed out that illness is experienced not simply as suffering but "as a suffering with a particular portent and meaning, as a suffering of a specific kind." He notes, for example, that a person with urethritis may experience the illness *as* "gonorrhea," or as "likely to be gonorrhea." A lump in the breast may be apprehended as "cancer" or "likely cancer." At this level the patient's experiencing of illness is influenced by the theoretical understandings that are embedded in the lifeworld. That is, for those who live in a highly technological society, "pathoanatomically based theoretical concepts are expressed in the constitution of the lived experience of one's body" (Engelhardt, 1982, p. 141). Consequently, individuals in such a society come to experience themselves not simply as having pain, or pain "in the chest," but as "having a heart attack." At the level of "disease" the patient assigns explanatory meaning to the experience of illness, although such meaning may be more or less sophisticated and does not coincide with the theoretical explanation of the "disease state" conceived by the physician.[53]

Cassell (1979, p. 211) makes the important point that at the level of "disease" the meaning that the patient assigns to experience is also influenced by its association with the experiences of significant others in the patient's life. For example, if I notice stiffness and pain in my fingers and my mother had arthritis, I am likely to apprehend such pain and stiffness *as* "arthritis." Likewise, an individual with a family history of heart disease may well experience chest pain *as* "having a heart attack," whereas another individual with no such history might dismiss such pain as merely indigestion. In this regard Cassell (1979, p. 203) distinguishes between the patient as experiencer and the patient as assigner of understandings. He argues that the patient assigns meaning to immediate pre-reflective sensory experience at two distinct levels: the first involves interpreting the sensation as, say, painful or dysfunctional; the second involves assigning the meaning of, say, "possible gallbladder disease" to the interpreted sensation of pain. These two levels of interpretation reflect the difference between "suffered illness" and "disease" identified by Sartre.

It is obvious that the manner in which the patient apprehends illness at the reflective level is a function of his or her biographical situation. That is, the assignment of meaning to pre-reflective sensory experience will be influenced by what Cassell (1979, p. 212) has called "the collectivity of (one's) meanings." Consequently, cultural meanings are an important determinant in the manner in which illness is apprehended. Engelhardt (1976, p. 262), for example, argues that the very determination of which functions are or are not proper to humans (i.e., the initial interpretation that the pre-reflective sensory experiencing is "alien" and thus a symptom of illness as opposed to something else) involves a value judgment. This value judgment may vary in different cultures. As Kleinman (1988) has shown, what is regarded as "natural" functioning depends upon the shared understandings of a particular social group. The meanings assigned to body sensations are a part of the shared common-sense knowledge of the group (in Schutz's terms the social "stock of knowledge-at-hand") so that there is a shared appreciation of what the sickness experience is. However, this shared stock of knowledge is not common to every culture or to each historical period. Kleinman (1988, p. 11) notes, for example, that even at the basic level of interpretation there are differences in the way that we talk about the immediately sensed pain of, say, headaches.[54] Indeed, studies have shown that what patients say, and their behavior in response to pain, is influenced by their ethnic group (Cassell, 1979, p. 203). (However, as Cassell points out, the immediate experience of pain is universal in the sense that it represents a similar kind of experience in all groups. What one group describes as "pain" another group does not describe as "itch," and so forth. That is, at the pre-reflective level the lived experience of pain has an inherent negative quality although the significance of that negativity depends upon how it is interpreted according to the cultural and personal meanings of the person who experiences it.)

As has been noted, the apprehension of illness as "disease" ("my abdominal pain is possibly gallbladder disease"; "I have a temperature so I must have a virus") is a direct reflection of the particular lifeworld in which the patient is situated. Not only is it the case that the patient's experiencing of illness is influenced by the theoretical understandings that are embedded in his or her particular lifeworld, but symptoms are interpreted according to special significances within the distinctive lifeworlds shaped by class, ethnicity, age and gender.[55] For example, Kleinman (1988, p. 24) notes that complaints associated with menopause are common among white middle-class women in mid-life, whereas women in other cultures have no conception of this life transition as an illness. Similarly, premenstrual tension (constituted as "PMS") is a constellation of symptoms of illness which is unheard of in much of the world. Meanings inherent in the lifeworld of the patient determine whether sensory experience at the pre-reflective level is apprehended as "suffered illness" and further as "disease."

## *Summary*

In sum, then, it is important to recognize that the meaning of illness is constituted by the patient at both the pre-reflective and reflective levels. The fundamental level is that of pre-reflective sensory experience. At this level one's immediate experience is such that it leads one to become aware of some disruption in the manner in which one "exists" one's body. That is, some unusual sensory experience (such as pain, weakness, or some visual apprehension of an alteration in body) causes the patient to shift attention from ongoing involvement in projects in the world and to focus upon the bodily disruption. Once the immediate experience of disruption is thematized at the reflective level, it may be apprehended as "suffered illness." "Suffered illness" is a synthetic totality in that it incorporates the immediate bodily sensations – the various and varied aches and pains – as parts of a larger whole. In particular, the unusual sensations are interpreted as symptoms which point to or characterize a more complex entity – illness. Furthermore, at this reflective level, the disruption is identified and located as, say, "in the leg" or "in my leg." It is important to note that both pre-reflective sensory experience and "suffered illness" represent lived experience. The patient lives through "suffered illness" *as* the collection of aches and pains which characterize the immediate sensory disruption at the pre-reflective level.

At a further interpretive level the patient apprehends "suffered illness" to be "disease." The lived body becomes objectified as a neurophysiological organism and the immediate sensory disruption is apprehended as a particular illness. The person who grasps illness as, say, "having a heart attack" is reflecting upon and assigning a specific meaning to the immediately sensed experience of pain, or rather pain "in the chest." The understanding of what it is to have a heart attack may represent more or less detailed knowledge depending upon whether or not the patient has discussed the illness with a physician. In any event, as Sartre points out, the patient's conceptualization of "disease" incorporates bits of knowledge acquired from others and, as such, it is quite different from pre-reflective sensory experience and "suffered illness."[56] At the level of "disease," illness is an object – a "being-for-others" – and, as such, it is transcendent to subjectivity and no longer represents the lived experience of illness.

3. THE PHYSICIAN'S APPREHENSION OF THE PATIENT'S ILLNESS

Consider now the meaning of the patient's illness for the physician. Sartre has identified this level of constitution of meaning as that of the "disease state" and he argues that, as a "disease state," illness is wholly conceived as "a question of bacteria or of lesions in tissue." Sartre's insight is, of course, that the conception of illness as a "disease state" is quite different from its apprehension in terms of pre-reflective sensory experience, "suffered illness" or "disease." As a "disease state" illness is thematized in terms of theoretical, scientific constructs. That is, the patient's immediate experience is wholly subsumed under the causal categories of natural scientific explanation.[57]

In western scientific medicine the prevailing model of illness is the biomedical model (Engel, 1977b; Baron, 1981; Schwartz and Wiggins, 1985; McWhinney, 1983; Kleinman, 1988). According to this scientific account illness is identified as a pathological or pathoanatomical fact.[58] As Engelhardt (1982, p.47; 1976, p. 260) notes at the level of the "disease state" illness is conceptualized according to the pathoanatomical, pathophysiological and microbiological nosologies of modern medicine (i.e., according to the nosologies of the basic sciences). So, for example, peptic ulcer disease is equated with an ulcer crater in the duodenum and with various aberrant complex pathophysiological and hormonal processes (Schwartz and Wiggins, 1988, p. 140). The patient's illness is thereby thematized *as* a pathological and pathophysiological process (i.e., *as* the anatomical fact of ulcer crater). Indeed, as has been noted in the previous chapter, many biomedical practitioners tend to assume that such "objective facts" alone constitute the reality of illness.[59] That is, it is concluded that patients' complaints that do not correlate with demonstrated pathoanatomical and pathophysiological findings are not *bona fide* illnesses.

That such findings take on a life of their own may be illustrated in the following example from my own experience. Some time subsequent to being diagnosed as having multiple sclerosis, I was hospitalized because of significant motor weakness and muscle pain. I was having great difficulty climbing stairs, walking more than a very short distance, and so forth. While the motor weakness was not atypical of MS, the muscle pain was unusual and my physician thought it necessary to investigate the possibility of a primary muscle disorder. Various tests were performed culminating in a muscle biopsy. The initial pathology report indicated that there was a primary myopathic process going on but there was no explanation as to the cause. Since there was no clearcut definition of the problem, it was also not clear what therapy might be instituted to correct it. I was extremely discouraged by my inability to get around, by the continuing pain, and by the apparent inconclusiveness of the tests. In frustration I commented that, since the biopsy did not indicate what the problem was, nor what to do about it, we seemed to have gained little by performing the procedure. My physician replied, "Oh, but we have! Now we *KNOW* something is wrong." For me, as a patient, to know that something was "wrong" was to be acutely aware of my bodily dysfunction and discomfort, and my inability to carry out the most mundane of activities. For the physician, to know that something was "wrong" was to have "objective" evidence in the form of an abnormal pathology report with respect to the muscle tissue removed from my thigh.

The level of the "disease state" represents a further objectification of illness over the previous levels of "suffered illness" and "disease." Symptoms are reinterpreted in terms of physical signs – the objectivity of, say, visible lesions. Physiological processes become translated into objective, quantified data – laboratory values, images, graphs, numbers, and so forth. Disease is constituted as an entity defined via medical categories. Since disease is categorized in the same way as other natural phenomena, it can be viewed independently from the person suffering from the disease (McWhinney, 1983).

Earlier I distinguished between the "natural" and the "naturalistic attitude." It will be recalled that the aim in the "naturalistic attitude" is to grasp the nature of "reality" and to describe such "reality" in terms of some objective description which will accurately characterize the "thing-in-itself" apart from one's experiencing of it. As Husserl (1970b, pp. 5–7) notes the aim in the "naturalistic attitude" is to establish what the world is *in fact* and thereby to arrive at scientific, objective truth (such objective truth being captured in terms of the quantifiable data of mathematical-physical science). The "naturalistic attitude" represents, therefore, a theoretical abstraction from prescientifically intuited nature; yet such abstraction comes to be viewed as alone disclosing the fundamental nature of things (Husserl, 1970b, pp. 48–53; Schwartz and Wiggins, 1988, p. 140).

The "disease state" is grounded in the "naturalistic attitude." The aim is to reclassify the patient's experience of illness in terms of the findings of the basic sciences. As Engelhardt (1989) points out, this aim of modern scientific medicine exemplifies the notion that the basic sciences are understood as "telling the real truth," as presenting what really is the case.[60] In addition, it represents the reduction of the clinical to the basic-scientific. The patient's subjective experience is reclassified *as* the pathoanatomical and pathophysiological "fact" and such experience has validity as illness only to the extent that it may be so construed.

It is important to emphasize that the "disease state" conceptualized by the physician is not identical with the "disease" which is apprehended by the patient. A concrete example may serve to illustrate this distinction. Suppose one has a neurological disorder: At the pre-reflective level the disorder is immediately experienced as a dragging of the leg which manifests itself in terms of the inability to climb the stairs without difficulty, a propensity for tripping up the curb, and so forth. At the reflective level, the dragging of the leg is apprehended as "suffered illness." It signifies or points to a more complex entity of which the dragging of the leg is but one part. Furthermore, it is experienced not simply as the inability to climb the stairs but as a disorder which is located "in the leg" or "in my leg." When the illness becomes further construed as "disease," the dragging of the leg is experienced as "a dragging of the leg which may indicate neurological disease" or as "possible multiple sclerosis" or "possible brain tumor." If a visit to the physician confirms, say, "multiple sclerosis" then from that point on one thematizes the dragging of the leg *as* "multiple sclerosis." Consequently, if asked how one is faring, one is likely to say "the multiple sclerosis is progressing" or "I'm having problems with the M.S." It is important to recognize, however, that even though, as a patient, one may understand one's "disease" *as* "multiple sclerosis" and, consequently, as involving a disruption of the nerve pathways which control motor functioning, one does not experience the disruption of the nerve pathways directly (i.e., one does not directly experience the lesion in the central nervous system which is the disease known by the physician).

In contrast, the physician construes the patient's illness directly as a disease state (i.e., *as* "bacteria and lesions in tissue"). It is not simply that the physician experiences the fundamental alien body sensation as "suffered illness" and further as "multiple sclerosis" (indeed since this represents the subjective experience of illness, the physician is unable to apprehend this at all) but rather that the physician regards the fundamental entity as *being* the lesion in the central nervous system. Thus, for the patient, the fundamental entity is the body painfully-lived whereas for the physician the fundamental entity is the disease state.

## 4. IMPLICATIONS FOR MEDICAL PRACTICE

The foregoing analysis reveals the enormous complexity of the meaning of illness. In particular, this analysis underscores the philosophical importance of the difference between meaning which is grounded in lived experience and meaning which represents an abstraction from lived experience. The "disease state" – as construed in the "naturalistic attitude" – represents a theoretical abstraction which is distinct from, and not identical with, the patient's experiencing. Illness in its complexity cannot be reduced to its conception as a pathoanatomical and pathophysiological fact. If one is to alleviate the patient's suffering, it is necessary to pay explicit attention not only to the patient's sensory experience of illness but also to the patient's apprehension of illness at the reflective level (such apprehension being a function of the patient's own particular meanings, evaluations, expectations, and so forth).

In this regard it is important to emphasize that there is a distinction between suffering and clinical distress. Suffering is experienced by persons, not merely by bodies (Cassell, 1982, pp. 639–645). Suffering occurs at the reflective level and is intimately related to the manner in which the patient apprehends the illness (i.e., the meaning and sig-nificances assigned to the pre-reflective sensory experience by the particular patient). So, for example, suffering may involve physical pain but is not limited to it. It is the particular significance accorded to such pain (or other physical anomaly) that causes suffering. Engelhardt (1989) asks us to consider, for example, the difference between pain recognized as a beginning heart attack signalling possible death (full of foreboding and danger), pain a runner feels during a race, pain experienced as a part of loveplay, pain experienced by a patient with terminal cancer (signalling the inevitability and proximity of death), and pain thought by the patient to be a sign of a heart attack but which instead is diagnosed as a non-threatening ailment. It is clear from these examples that not only will certain of these instances of pain *not* be construed as "illness" by the patient but, in addition, the meaning that the patient assigns to the pain will determine whether or not such pain involves suffering.

As has been noted, it is the case that meanings inherent in the life narrative of the patient determine whether or not pre-reflective sensory experience is apprehended as "suffered illness" and "disease." It is also the case that such meanings determine whether "disease" involves suffering. That is, the significance of alien sensations (such as stiffness in the joints, a disfigurement such as an unsightly rash) will vary according to the particular patient's life situation. Whereas stiffness in the joints (grasped as "arthritis") may be simply a nuisance to one individual, it might represent untold suffering to a professional concert pianist. This is simply to note that the manner in which illness is apprehended and the suffering which accompanies illness, is integrally related to the whole pattern of a person's life.[61]

Cultural definitions of illness can also be a source of suffering to the sick person. Such definitions influence the behavior of others towards the person who is ill and the behavior of the sick towards themselves. Cultural norms and social rules determine whether or not the sick will be considered foul or acceptable, whether they are to be pitied or censured, and whether or not they should be isolated (Cassell, 1982, p. 642). For example, it might be the case that a disfigurement which causes no overt loss of function or sensory distress leads to suffering at the reflective level due to the negative reaction of others who find the disfigurement unacceptable or unattractive. Similarly, the suffering which accompanies disability is not solely caused by the actual loss of function but incorporates the patient's recognition of a devaluation in status which reflects cultural values.[62]

Since suffering is intimately related to the manner in which illness is apprehended by the patient, it is clear that the alleviation of suffering requires that attention be paid to the patient's meanings. Since suffering is not identical with clinical distress (and illness is not identical with the "disease state"), suffering is not necessarily alleviated when attention is given solely to illness as a "disease state." Indeed, Cassell (1982, pp. 639–645) argues that the failure to understand the nature of suffering can result in medical intervention which (though technically adequate) not only fails to relieve suffering but itself becomes a source of suffering for the patient. An obvious example would be the suffering engendered by the mutilation of one's body following a radical mastectomy or the distress accompanying impotence following a radical prostatectomy – both procedures, however, representing successful intervention in terms of cancer therapy.

It is important to note that only the patient can gauge whether or not medical intervention is sufficiently disruptive to cause personal suffering.

Whereas the physician may consider the side effects of a particular medication to be relatively innocuous, the patient may find such side effects unacceptable in terms of the resulting disruption of daily living in the context of his or her particular life narrative. This is the case even in the event that the treatment is not for a serious or life threatening illness. For example, if I am a long distance truck driver or a college student studying for finals, I may have problems if I need to take any medication which causes drowsiness. Will such treatment cause me to suffer? Not necessarily. I may be able to so modify my activity as to compensate for the detrimental effects of the treatment (or I may be able to substitute another type of treatment). But suffering *could* occur if the treatment disrupts my activities on an ongoing basis and if those activities are of particular significance to me.

Changes in meaning may relieve or exacerbate suffering. In this connection it is important to consider the impact of diagnosis on the patient's apprehension of illness. In the example provided by Engelhardt (see above), the diagnosis changed the meaning of the experience of pain when what was initially thought to be pain associated with an impending heart attack was diagnosed instead as a non-threatening ailment. Diagnoses are, however, themselves permeated with personal and cultural meanings. Indeed, whether or not a diagnosis has validity as a *bona fide* illness is in large part a function of the culture and the historical period. For instance, Kleinman (1988, pp. 102 ff.) indicates that "neurasthenia" is no longer a "fashionable" diagnosis in North America (although at one time it was considered the "American disease"). Consequently, it has been banished from orthodox nosology in this country although it is considered a legitimate physical ailment in China.[63] Susan Sontag (1978; 1988) has explored the cultural meanings (and stigma) associated with such diseases as cancer and AIDS. Others (Herzlich and Pierret, 1987) have similarly noted that each society structures the meaning of illness in very different ways according to its own values. This collective representation of illness enters into the meaning which the diagnosis has for the patient. For example, Ronald Carson (1986, pp. 48–50) emphasizes that cancer is loaded with symbolic meaning, "surrounded by metaphoric meanings of a particularly horrid kind." Consequently, facts about cancer are "loaded facts, freighted with a significance that, if undiscerned or unacknowledged, will likely thwart even the most well-intentioned dedication to patient care" (Carson, 1986, p. 50).[64]

Since changes in meaning can alleviate or exacerbate suffering, it is important to understand the meaning a diagnosis has for a particular patient. The only way to find out is to ask the patient. In this way it may well be possible for the physician to alleviate unnecessary suffering resulting from evaluations and expectations which might be interpreted differently. For example, I well remember receiving the diagnosis that I had multiple sclerosis. A few days earlier I had fortuitously read an article in a popular magazine in the doctor's office which related the story of a young woman (a former beauty queen), stricken with M.S., who was now severely disabled and confined to a wheelchair. Consequently, on hearing the diagnosis, my first question was, "Will I end up in a wheelchair?" The physician replied that unfortunately he could give me no guarantees for the future. Not unnaturally I interpreted this response to mean that I would *indeed* become disabled and perhaps in the near future. (This interpretation was reinforced when, on arriving home, I looked up "multiple sclerosis" in an outdated edition of an encyclopedia. The entry stated that M.S. was an incurable, progressive disease of the central nervous system culminating in total paralysis and death.) While the doctor's response to my question was not incorrect (he couldn't guarantee my future physical state – for that matter who can?), it is certainly the case that not *all* M.S. patients end up severely incapacitated and my initial dire interpretation of my situation would have been less distressing had he included such information in his response to my question. The point is that, if the physician is sensitive to the patient's interpretive understanding of illness, he or she can act as an arbiter of meaning – perhaps enabling the patient to modify or change an inappropriate interpretation of the situation.[65]

With regard to the impact of diagnosis, the failure to construe illness as a "disease state" may also be a source of suffering to the patient. As has been noted the biomedical model requires that the patient's complaints be correlated with demonstrated pathoanatomical or pathophysiological findings if such complaints are to be recognized as *bona fide* illnesses. Thus, a scientific diagnosis validates the patient's experience and the lack of such a diagnosis suggests such experience is not to be taken seriously as a medical problem.[66] Engelhardt (1982, p. 50) shows the distrust of nonpathoanatomically or nonpathophysiologically based complaints has become integral to the character of modern western medicine. In medical training (most of which takes place in high technology hospitals) patients with recurring but nonspecific complaints which are not easily assimilable within the scientific model are regularly referred to as "crocks" and their symptoms dismissed as unimportant (Donnelly, 1986; Konner, 1987). Yet patients seeking care for complaints which are not easily referred to specific pathoanatomical or pathophysiological lesions are numerous in general practice. For such a patient to have the experience of illness dismissed as not "really" illness, to be told that it is "all in your head" or "there is nothing wrong with you" is a source of dismay (especially as this assessment not only contradicts actual experience but implies that the distress is not legitimate).[67]

There are other reasons why it is important for the physician to have some understanding of the patient's apprehension of illness. In particular, such an understanding enables the physician to be a better therapist. It is, for example, essential to recognize that the patient's conception of "disease" is not identical with the naturalistic account of the "disease state." George Engel (1987, pp. 107–109) shows that ignoring the patient's meanings can result in errors in diagnosis and therapy. He details a case where the physician tacitly assumed that the patient's description of his complaint as "spitting blood" meant hemoptysis (the coughing up of blood). This unquestioned assumption led to thorough pulmonary studies including bronchoscopy, none of which yielded an explanation for the bleeding. However, when subsequently asked exactly what he meant by "spitting blood," the patient proceeded to describe in detail feeling something flowing down the back of his throat which when "hawked up" proved to be bloody mucus. He had never "coughed up" blood. His problem proved to be a small varix oozing blood in the nasopharynx which was subsequently cauterized. Cassell (1979, p. 210) notes that patients often describe their symptoms in disease terms (e.g., "I had a virus last week"). However, it is important for the physician to recognize that the patient's conception of, say, "virus" is significantly different from the doctor's conception and such differences need to be carefully explored.

The phenomenological analysis of the meaning of illness reveals that at the level of "disease" the patient assigns explanatory meaning to the experience. In this connection Kleinman (1988, pp. 121–122) argues that it is vital for the physician to understand the patient's explanatory model of illness. Explanatory models are notions that patients, families and practitioners have about specific illness episodes. The patient's explanatory model involves answers to such questions as: what is the nature of the problem, why has it affected me, why now, what course will it follow, what treatment do I desire, what do I fear most about the illness and its treatment, and so forth. Obviously, the patient's explanatory model will differ from the physician's explanatory model. Kleinman (1988, pp. 121–122) notes that the elicitation of patient and family explanatory models aids the practioner in organizing strategies for clinical care. The doctor who does not sufficiently take account of the patient's understanding of "disease" may miss crucial features of the illness or may prescribe inappropriate treatment.

Furthermore, it is important that the practitioner take the time effectively to communicate to the patient his or her explanatory model of the illness. Failure to ensure that the patient clearly incorporates this model may result in appropriate treatment being ignored. Engelhardt (1982) argues, for example, that until patients see themselves as "diabetics" or "hypertensives," they do not regularly do the things that diabetics or hypertensives ought to do. Engelhardt here makes the important point that patients can, in fact, learn to interpret their illnesses in terms of the objective, quantitative data that characterize the "disease state." For example, in living with leukemia, Stewart Alsop (1973) eventually came to structure his illness in terms of platelet counts. Chronically ill individuals, such as diabetics or sufferers from chronic renal insufficiency treated by kidney machines, must closely follow their condition and may perceive their illness in terms of levels of glycemia, measurements of arterial pressure, phosphorus-calcium ratios, and so forth (Herzlich and Pierret, 1987, pp. 94–97).[68] (Of course, this interpretation represents an abstraction from their lived experience. Alsop, for instance, found that his actual experience – how he felt – did not necessarily correlate with the quantitative data provided by the platelet counts, although the results of such counts naturally affected how he responded in terms of the threat posed by his illness.)[69]

An important factor which contributes to the unshareability aspect of illness is the incommensurability of inner and outer time. As I noted earlier, the patient lives through his or her illness in inner time (lived time), yet must describe the illness in terms of outer time (objective time). The physician, on the other hand, uses the objective time scale to measure the physical events and biological processes which define the patient's illness as a disease state. In light of the foregoing analysis of pre-reflective sensory experiencing, "suffered illness," "disease," and the "disease state," it is possible to give a more detailed account of the manner in which patient and physician constitute the temporality of illness differently.

At the levels of pre-reflective sensory experience and "suffered illness" the patient lives through his or her illness in inner time. A sensory experience such as pain is not measured by the sufferer according to the units of the objective time scale. Rather illness-as-lived is experienced as an ever-present consciousness of disorder. Pains, for example, are experienced in the ongoing flow of inner time (i.e., as an integral part of a continuum – a continuum which incorporates not only the present now-point but those now-points which are just past, as well as future now-points which are to come). Similarly "suffered illness" (the psychic entity, pain-as-object) is experienced in its immediacy in inner time. The pain "in the stomach" which is just-past is retained in retentional consciousness, just as the future pain "in the stomach" is anticipated (protended), as a part of the present consciousness of disorder.

At the level of "disease," however, the patient (in objectifying the body and in assigning explanatory meaning to the experience of illness) reflects upon, rather than lives through, his or her illness. Moreover, in describing "disease" to the physician, patients attempt to do so by providing a narrative history. In so doing, they are obliged to recollect past events and relive illness, not in its immediacy but in an "as-if" presentation (i.e., one recalls past pains rather than living through them).[70] Thus, the temporality of "disease" differs markedly from the temporality of pre-reflective sensory experience or "suffered illness." In reflecting on "disease" and providing a narrative history, patients often present a sequence of events.

The first thing that happened was that my eyes swelled up ... the very next day ... I got a very bad pain in the back of my left leg.... And then that developed ... that lasted for about two days and then ... the whole procedure has been one thing after another (Cassell, 1985a, p. 13).

Such a sequence of events is reported according to the objective time scale which provides a common language for time.

In apprehending illness as "disease," and in giving a history, patients may also do so in terms of a causal chain, "where one event follows another, as though the first caused the second, caused the third, ad seriatum" (Cassell, 1985a, p. 31). The causal chain is constituted with reference to objective, rather than inner, time. Unlike pain and "suffered illness," which are experienced as a continuum of retentional and protentional phases in inner time, "disease" is reflectively described as a series of discrete, atomic instants which occur along a time line.

At the level of the "disease state" the physician apprehends the patient's illness as a temporal process wholly according to the units of the objective time scale (i.e., as a causal process occurring through time). It should be noted, however, that although the physician construes the "disease state" in terms of a causal chain, the events identified in this chain may not (and almost certainly do not) coincide with those events identified by the patient.[71]

An explicit awareness of the temporal dimension of illness is important for creating a shared world of meaning between physician and patient. As has been noted, it is often hard for patients to measure their immediate experience of disorder in terms of the objective time scale (since they are living through such disorder in inner time). Such difficulty in communicating illness should be recognized as a function of temporal experiencing rather than as an indicator that the patient is an unreliable narrator of his or her illness story.

As Sartre has pointed out, the objectification of illness as "disease" (i.e., as a "being-for-others") results in a sense of alienation from one's body. The analysis of temporality further suggests that a contributory factor to this sense of alienation is the difference in temporal constitution between the various levels of intentional experience. As one moves from the temporality of inner experiencing at the level of lived experience to a reflective description in terms of objective time, illness is transformed into an objective entity which is transcendent to subjective consciousness. The further one moves from lived experience, the greater is the sense of alienation from one's body.

In considering the manner in which the patient apprehends illness, it has been noted that illness engenders a fundamental change in the manner in which the body is experienced. Abnormal sensory experience renders the body thematic. Rather than living his or her body unreflectively, the patient focuses on the bodily disruption and attempts to discover its meaning. Furthermore, at the level of "disease" the patient objectifies the lived body and apprehends it to be a neurophysiological organism. This objectification of the body results in a sense of alienation between body and self which is intrinsic to the experience of illness. Such alienation is intensified as a result of the physician's construal of the body as a scientific object within the "naturalistic attitude." As is the case with illness, the manner in which the body is apprehended differently by physician and patient reflects the distinction between meaning which is grounded in lived experience and meaning which is not so grounded. This bodily constitution will now be explored.

# CHAPTER THREE

# THE BODY

In considering the manner in which the body is apprehended by the patient in illness, it is important first to explore the manner in which the body is apprehended in normal circumstances. In particular, one should distinguish the lived body (the body experienced at the pre-reflective level in a non-objective way) and the objective or physiological body (the body apprehended at the reflective level as a material objective entity among other entities within the world).

## 1. THE LIVED BODY

The phenomenological analysis of body provided by Sartre and Maurice Merleau-Ponty reveals a fundamental distinction between the lived body (the body as it is immediately experienced in a non-reflective or pre-reflective manner) and the objective or physiological body. In particular, such an analysis discloses that at the pre-reflective level (i) the body is not explicitly thematized *as* body (i.e., it is not apprehended as a physiological body or as a material object among other material entities within the world); (ii) the relation with lived body is an existential, rather than an objective, relation. At the level of lived body I do not "have" or "possess" a body, I *am* my body; (iii) there is thus a fundamental identification with body at the pre-reflective level such that there is no perceived separation between body and self; and (iv) the lived body exhibits certain features which are essential to embodiment. Such features include being-in-the-world, bodily intentionality, primary meaning, contextural organization, body image, and gestural display and significance.[72]

At the pre-reflective level, as Sartre (1956a, pp. 401–402) notes, the body is not explicitly thematized *as* body. In the normal course of events I do not experience my lived body as a biological organism (i.e., as a brain, skeleton, nerve endings, and so forth). My lived body is essentially that which is perpetually "forgotten" or "surpassed" in carrying out my projects in the world (Sartre, 1956a, pp. 429–430). In writing a letter, for example, I am not explicitly aware of the neurophysiological mechanism

51

which controls the movement of my hand and the grasping of the pen. Indeed, I am not even conscious of my hand at all. My attention is wholly directed to the task at hand. While the lived body is present in every action, it is "invisible." The act reveals the writing of the letter, not the hand which writes. While the lived body is always present, always the center of reference for my world (and in that sense always the "referred to" of my world), it is the "inapprehensible given" – a center which is indicated but never grasped as such (Sartre, 1956a, pp. 425–427).

What Sartre is concerned to emphasize here is that at the level of lived body any consciousness of the body is a non-thetic consciousness. So, for example, at the level of lived experience pain in the eyes is simply exhibited as the eyes "painfully-lived." Since I *am* my body, in that I am an embodied subject, it takes an act of reflection to make my body stand out *as* body (i.e., to turn my lived body into an object for me-as-subject).[73]

My lived body is both the total center of reference which things indicate and the instrument and end of my actions (i.e., the center of a complex system of instrumentality and the reference of the series of instrumental acts). Nevertheless, since I *am* my body, I do not perceive it to be an instrument like other instruments:

I am not in relation to my hand in the same utilizing attitude as I am in relation to the pen; I am my hand.... I can apprehend it – at least in so far as it is acting – only as the perpetual, evanescent reference of the whole series ... my hand has vanished; it is lost in the complex system of instrumentality in order that the system may exist. It is simply the meaning and orientation of the system (Sartre, 1956a, p. 426).

As an embodied subject I find myself always within the world in the midst of environing things. I am "embodied" in the sense not that I "possess" a body but in the sense that I *am* my body. Rather than being an object *of* the world, my body is my particular point of view *on* the world (Merleau-Ponty, 1962, pp. 70, 90).[74] Indeed, as Merleau-Ponty (1962, p. 92) notes, it is by means of my body that I have access to the world in the first place. Sensory experience is, after all, the sole means by which the environing world of things is at all disclosed to me. As my orientational locus in the world, my body both orients me to the world around me by means of my senses, and positions the environing world in accordance with my bodily placement and actions.[75] From the point of view of my experience of the world, to perceive something is necessarily to be related to it by means of my body.

Furthermore, the lived body exhibits certain features which are essential to embodiment. Such features include being-in-the-world, bodily intentionality, primary meaning, contextural organization, body image and gestural display.

## Being-in-the-World

My bodily engagement in the world is an active one. Rather than being an exclusively physical thing devoid of intentionality, the lived body is an embodied consciousness which engages and is engaged in the surrounding world.[76] Not only do I constantly find myself within the world but I continually move towards the world and organize it in terms of projects, and so forth. In this respect, says Merleau-Ponty (1962), sensory perception is neither a purely mechanical, physiological process nor, alternatively, a purely psychological one. Rather, sensing exhibits a "bodily intelligence and affectivity." Thus, the function of the lived body can only be understood insofar as the lived body is a being-in-the-world. It is the global presence of the situation which gives meaning to the sensory stimuli and "causes them to acquire importance, value or existence for the organism" (Merleau-Ponty, 1962, p. 79). Perception cannot be divorced from the concrete situation of the perceiver. Every sensible quality not only exists within a specific milieu but is determined and defined with respect to the "task at hand" (Zaner, 1964, p. 159). Consequently, bodily acts must be understood in terms of their being acts which take place within a certain situation having a certain practical significance for the embodied subject.[77]

## Bodily Intentionality

As a practical field of significance, the world arouses in the lived body certain habitual intentions (for example, manipulatory movements such as grasping and so forth). Consequently, objects are apprehended as manipulatable or utilizable by the body (Merleau-Ponty, 1962, pp. 81–82).[78] In its directedness towards (attentiveness to) the world the body thus exhibits a bodily intentionality. The parts of the body may be understood as "intentional threads" linking it to the objects (the world) which surround it. As such, objects are "poles of action" which delimit a certain situation and which call for a certain mode of resolution, a certain kind of work (Merleau-Ponty, 1962, p. 106). Perception reveals objects as "invitations" to my body's possible actions on them, as "problems" to my body (Zaner, 1964, p. 131). Consequently, every perceived object is inseparably connected to my body since my body is the locus of all intentions. The surrounding world is always grasped in terms of a concrete situation. Objects are encountered, for example, as the cloth "to be cut up," the book "to be read" or "to be replaced on the shelf." Bodily space is given as an intention to take hold – a matrix of bodily action.

Embodied consciousness is, then, in the first place not a matter of "I *think*" but of "I *can*." In the action of the hand which reaches for the pen is contained a reference to the object, not as object represented, but as that highly specific thing at which I aim "in order to" effect some action. Every formula of movement presents itself to the body as a practical possibility, a sphere of action (Merleau-Ponty, 1962, p. 138). Consequently, objects are oriented as the context for the body's possible action. The object is presented to the body as a question, a problem to be solved (Zaner, 1964, p. 177). Thus, into every geographical setting is built a behavioral one – a system of meanings by means of which the individual organizes the given world (Merleau-Ponty, 1962, p. 112). The embodying organism is experienced as "always in the midst of environing things, in this or that situation of action, positioned and positioning relative to some task at hand" (Zaner, 1981, p. 97).

### Primary Meaning

Merleau-Ponty (1962, p. 132) argues, moreover, that the primary imposition of meaning is that afforded by the body through what he calls "physiognomic perception." As the work of such developmental psychologists as Jean Piaget (1970) has demonstrated, the infant first understands the world through the experiences of sense perception and bodily action, and only subsequently through the development of rational and conceptual thought. There is thus a primary "knowing" which is a "knowing" through the body. To be attentive to things is to exist towards them in a manner which "precedes essentially all thematization, categorization and predication" (Zaner, 1964, p. 188). The meaning afforded by sensory-motor experience is a direct response to the world and is prior to any act of reflection or conceptualization.

Furthermore, the primary perceptual relation between body and object is that of "form-giving" (Zaner, 1964, p. 154). Sensory perception is already charged with meaning in that the object is always grasped as a significant whole against a background of co-perceived things (i.e., in a figure/ground relation). As Merleau-Ponty (1962, pp. 3 ff.) points out, we do not experience the world in terms of pure, isolated, sensations (i.e., in terms of such sense data as meaningless red patches). In our actual common-sense experiencing of the world there is no such thing as a pure impression or isolated datum of perception. "To perceive is not to experience a host of impressions ... it is to see, standing forth from a cluster of data, an immanent significance" (Merleau-Ponty, 1962, p. 22).

Furthermore, implicit in any concrete situation is a set of meanings "whose reciprocities, relationships and involvements do not require to be made explicit in order to be exploited" (Merleau-Ponty, 1962, p. 129). When I walk across the room, for example, I know without thinking about it that I must walk around the chair in order to get to the door. Locations and perceptions are immediately apprehended in relation to my bodily placement without being made explicit. Beneath objective space is a primitive spatiality of the body.

### Contextural Organization

Zaner (1981, p. 45) notes that not only does perception disclose the world in terms of a figure/ground relation but the bodily organism itself exhibits a figure/ground relation. To execute any movement such as raising one's arm (figure) requires a definite background attitude for the rest of the body (ground). Any change in movement results in a change in background attitude. Indeed, every bodily performance (sensory, motor, emotive) is necessarily and intrinsically implicated with others. This leads Zaner (1981, pp. 92–109) to suggest that the body may be considered as a complex, organized "contexture." That is, as a systematic totality of intrinsic references or functional significances, a "whole" composed of interrelating "parts" and "members." Indeed, it is a fundamental feature of embodiment that it is a part/whole, a "unity-in-difference" phenomenon at every level (Zaner, 1981, pp. 62–63). Moreover, Zaner (1981, p. 107) argues that body, consciousness and world unite to form a unique and complex whole, a "complexure," whose "parts" are themselves strictly inseparable, albeit distinguishable, contextures.

## Body Image

The lived body is experienced as an integrated system of coordinated body movements which are distributed spontaneously among the various body segments. For the most part the individual does not consciously effect this coordination. As Merleau-Ponty (1962, p. 149) notes, "I do not bring together one by one the parts of my body; this translation and this unification are performed once and for all within me; they *are* my body itself." When I reach for the glass, for example, my body coordinates not only the physical movements of the arm but also links tactile and visual sensations.

Furthermore, I do not behold the relations between the parts of my body as a spectator. Rather, I know where my limbs are through a "body image" in which all are included. My body image is the total awareness of my posture in the intersensory world, a "form." This "form" is dynamic. My body appears to me as "an attitude directed towards a certain existing or possible task" (Merleau-Ponty, 1962, p. 100).[79] In addition, the body is experienced not only in terms of its present set of positions but also as "an open system of an infinite number of equivalent positions directed to other ends" (Merleau-Ponty, 1962, p. 141). This system of equivalents, this "immediately given invariant whereby the different motor tasks are instantaneously transferable" is included in the body image. It follows then that body image is not only an experience of my body but an experience of my body-in-the-world.

## Gestural Display

Just as it is through my body that I perceive things, so it is through my body that I understand the bodily actions of others. When my neighbor waves to me from across the street, I understand his gesture not through some act of intellectual interpretation but in a sort of "blind recognition." The meaning of the gesture is understood through my own body's capacity to express itself in gestures:[80]

The communication or comprehension of gestures comes about through the reciprocity of my intentions and the gestures of others, of my gestures and intentions discernible in the conduct of other people. It is as if the other person's intention inhabited my body and mine his (Merleau-Ponty, 1962, p. 185).

In this connection it is important to note that body gesturing is not, in its primordial appearances, a matter of some sort of "internal" goings-on being "pressed-outward" by bodily movements and attitudes (Zaner, 1981, p. 63). Frowns and redness in the face do not simply express anger. They *are* the anger.[81]

It should also be noted that when I perceive the Other, I perceive his body as a totality and I perceive him always as being in a situation. Sartre (1956a, p. 455) argues that in isolation the gesture would mean nothing. I do not perceive, by itself, a "clenched fist." I perceive a man who in a certain situation clenches his fist. It is this totality, "body in situation," which is anger. Moreover, when my friend waves to me, I do not see an arm raised against a motionless body. I see my friend-who-raises-her-arm "in order to" – in order to attract my attention, express her friendliness, and so forth. I perceive her body as a totality and her gesture as having practical significance.

Zaner (1981, pp. 63–66) suggests that the body itself (stance, forms and motions) is a gestural display, its various and varying configurations communicating meaning to others. This is perhaps what we refer to when we speak of "body language," a language which at times communicates more eloquently than speech. In addition, Merleau-Ponty (1962, p. 150) has noted that we develop a certain corporeal style, a certain bodily bearing which identifies the lived body as mine.

*Summary*

In summary, then, at the pre-reflective level of immediate experience the body is apprehended in the natural attitude as lived body. At this level I do not explicitly thematize my body *as* a body. Rather than being an object for me-as-subject, the body-as-it-is-lived (the lived body) represents not only my particular point-of-view on the world but my unique being-in-the-world. As such it is the locus of my intentions and the instrument of my active engagement in and with the surrounding world. Consequently, I do not simply "have" or "possess" a body. As an embodied subject I *AM* my body. There is, thus, a fundamental identification with the lived body such that there is no perceived separation between body and self.

Moreover, certain essential features such as being-in-the-world, bodily intentionality, primary meaning, contextural organization, body image and gestural display are characteristic of the lived body. Such features must be taken into account when examining the manner in which the patient experiences the body at the pre-reflective level.


## 2. BODY AS OBJECT

As has been noted, at the pre-reflective level I do not explicitly thematize my body. Rather than being the object of attention, it is that which is surpassed in carrying out my projects in the world. To grasp my lived body *as* body requires an act of reflection which necessarily transforms it into an object-body.

For Sartre (1956a, pp. 445–460) the apprehension of one's body-as-object – the awareness of one's own "thingness" – is revealed in the experience of "being-for-the-Other." Sartre (1956a, p. 461) argues that I first experience my body-as-object in the gaze of the Other. In the experience of being looked-at, I recognize not only my being-an-object for another subject but also the brute fact of my being as material, "physico-biological stuff." That is, I am aware that when the Other looks at me what he sees is a physical body (one that is engaged in certain activities within the world). Consequently, in apprehending myself as a "being-for-the-Other," I apprehend my body as an object and, further-more, as a physical body (in Sartre's terms as an "ensemble of sense organs," as "flesh"). I see myself through the eyes of the Other and I recognize my facticity.

In this regard, Sartre argues that my "being-for-the-Other" is synonymous with the Other's being-for-me. That is, when I observe my friend across the room, what I see is her physical body. When she waves at me, I observe the movement of her arm and I notice the manner in which she raises it to attract my attention, and so forth. In this manner I can "know" my friend's body in a way that she cannot. In living her body unreflectively, it is that which is surpassed in carrying out her projects in the world. She is not explicitly aware of the manner in which she raises her arm to attract my attention whereas I have an explicit awareness of her movements in that her body is an object for me.

The experience of "being-for-the-Other" is one of alienation. The body appears as Other-than-me, as a thing outside my subjectivity. For example, when I perceive the doctor's ear listening to my heartbeat, I experience myself as an object. The lived-body becomes designated as a thing outside my subjectivity in the midst of a world which is not mine.[82]

Merleau-Ponty (1962, pp. 194–195) argues that there is, thus, a fundamental "ambiguity" in the structure of lived body. While the lived body is that which is most intimately "me" and "mine" (the "own-body" which I *am)*, it is yet an object for others – being at once the "expression" and the "expressed" of my existence.

The apprehension of one's body as an object does not, however, arise solely in the experience of being an object-for-the-Other.[83] The body becomes present to consciousness as a material object in such mundane experiences as fatigue (when one is "dead" tired and must drag one's body around), stubbing one's toe against the corner of the bed and feeling the pain in the toe, participating in strenuous exercise and becoming aware of the rapid beating of one's heart, having one's arm go to sleep and experiencing it as a heavy, lifeless "thing," and so forth.[84] Gallagher (1986, pp. 148–149) notes that according to most researchers the body suddenly appears in the field of consciousness "when the organism loses or changes its rapport with the environment," i.e., in certain "limit situations" such as sickness or pain or in such positive experiences as sport, dance, and sexual excitement. In such situations the lived body is recognized as essentially corporeal – that is, it is apprehended as a physical, material entity.

In addition, Engelhardt (1973, p. 38) points out that the body presents itself as a mechanical, physical object in such everyday occurrences as brushing against poison ivy and breaking out in a nasty looking rash, drinking a large quantity of beer and noticing a corresponding rise in urine output, or sitting in the sun and seeing a resulting bleaching of one's hair. In such occurrences the body appears as a physico-biological "thing" among other objects to be felt, seen and acted upon. Moreover, the body is apprehended as something which is Other-than-me. I am not (nor could I be) personally involved in the various and varied physical processes which characterize my body as a neurophysiological organism.

It should be noted that, as is the case with illness, the manner in which one apprehends one's body as a physical entity will reflect the particular lifeworld in which one is situated. Thus, for those who live in a highly technological society the body will be apprehended as a physico-biological thing according to pathoanatomically based theoretical concepts. For example, if I have a cramp in my leg I will be cognizant of my body as being a "skeletal body with a neuromuscular system"; if I cut my finger and it bleeds, I will recognize it to be a "body with a circulatory system."[85]

Zaner (1981, pp. 48–55) suggests that to experience one's own body as Other is to experience the own-body as "uncanny" – the "uncanny" being something hidden (repressed) which suddenly makes its appearance. He argues that there are four senses in which the body is experienced as "uncanny": (i) the inescapable/the limitation; (ii) chill and implicatedness; (iii) hidden presence; and (iv) alien presence.

In the first case a sense of inescapability and limitation are essential to embodiment. While it is *inescapable* that I be embodied, it is a matter of contingency that I have this particular embodiment (i.e., this particular neurophysiological makeup). And this particular embodiment carries with it certain radical *limitations*. "In critical ways ... and whether I like it or not, there are some activities, postures, gestures, sensory encounters, and sensory refinements, etc., which are just not within my bodily scope, thanks to my being embodied by this and not some other body" (Zaner, 1981, p. 51). Thus, I must come to terms with the limitations of my embodiment. I am not free to do whatever I will. I must necessarily take my body into account. For instance, it is unlikely that I will become a professional basketball player if I am four foot eleven inches tall, or skilled in microsurgery if I lack coordination.[86]

Moreover, while my body is experienced as intimately "mine," there is a sense in which I belong to it, in which I am at its disposal or mercy:

My body, like the world in which I live, has its own nature, functions, structures, and biological conditions; since it embodies me, I thus *experience myself as implicated* by my body.... I am exposed to whatever can influence, threaten, inhibit, alter, or benefit my biological organism (Zaner, 1981, p. 52).

Since whatever happens to my body affects me, bodily experiences are experiences of *"corporeal implicatedness"*. I find myself to be that person who is bound to this particular embodiment and who is irrevocably bound to suffer whatever this particular body suffers. The recognition of this corporeal implicatedness (especially in its most radical form, that of my own going-to-die) may be accompanied by a sense of chill or dread. Thus embodiment is experienced not only as that which is most intimate –

"mine" – but as that which is "dreadfully and chillingly implicative" (Zaner, 1981, pp. 52–53).

In addition, the body is experienced as a *"hidden presence"* in that, as biological organism, it includes events, processes and structures over which I have no control and of which I have no awareness. In a sense my body seems to "carry on" without me and to have no need of me. Although I have an intimate knowledge of my body, yet I do not "know" it in fundamental ways. Even if I study anatomy and physiology, I do not directly encounter the "hidden presence" of *my* body. Rather I learn about *"the"* heart, *"the"* lungs, *"the"* metabolism (Zaner, 1981, p. 53). Rarely are these directly experienced or experienceable in my own case. In this sense, says Gallagher (1986, pp. 154–155), the body is "experientially absent." While physiological processes are lived through, they are not consciously lived. Most processes or happenings in the body which may be described in neurophysiology are unfelt and do not seem "to me" to be happening in "my" body.

Finally, my body manifests itself as other in the form of an essentially *"alien presence"* which has its own nature, its own biological rhythms, and so forth. "Whatever I want, wish or plan for, I irrevocably 'grow older,' 'become tired,' 'feel ill,' 'am energetic' – and these, at times, whether or not I plan my life, or my day even, so as to gain some control over my bodily 'moods'" (Zaner, 1981, p. 54). I am responsible for it, yet at its disposal, and at the same time it expresses and embodies me. My body is at once the most intimate yet alien presence.

It is clear that the objectification of the body at the reflective level involves a disruption of the unity of lived body. That is, as an object, the body may for the first time be apprehended as separate from the self. At the level of lived body I do not have an explicit awareness of my body as a separate entity. Rather I AM my body. I exist it. Therefore, the lived body is not a thing which "I" own and which makes me the subject and it the object.[87] Although there is an implicit, pre-reflective awareness of bodily identification at the level of lived body – for example, if I move my hand towards something, in some sense I "know" that it is "my" arm that moves – nevertheless, this is a non-objectifying experience. The body is not thematic to consciousness as a thing apart from the self.

At the reflective level the body becomes an object for me as subject. I now explicitly recognize this body as "my" body in a sort of "owned" recognition. Rather than simply existing my body, I am aware that "I" *have* or *possess* a body. It is, after all, "my" body as opposed to the body of my friend, Fred (a fact in which I may take pride if I have spent long hours at the gym "building it up," or for which I may have fleeting regrets as I view his svelte frame contrasted with my own rotundity). This is, to be sure, a unique kind of possession. I cannot, for example, separate from my body in toto (although I can rid myself of parts of it, such as an inflamed appendix). Consequently, there is – at the level of body-as-object – a sense not only of possession but also of identification. I understand myself to be conjoined with "my" body in a symbiotic relationship.

At the same time – in recognizing it as a material, physical object – I apprehend the body as Other-than-me. The apprehension of otherness of body is not necessarily a negative one. In positive experiences such as sexual arousal, sport, dance, and so forth, I may take delight in the physical nature of my body. However, as Zaner has pointed out, the apprehension of otherness of body may alternatively bring about a sense of alienation and "uncanniness." Under normal circumstances this sense of bodily alienation and "uncanniness" is, for the most part, a fleeting experience – one that is easily forgotten and passed over. As we shall see in the following analysis, however, in illness the apprehension of otherness of body is both negative and profoundly alienating.

### 3. LIVED BODY IN ILLNESS

Having examined the manner in which the body is experienced at the pre-reflective level under normal circumstances, I shall now explore the way in which the lived body is experienced when one is ill. In particular, it will be noted that illness strikes at the fundamental features of embodiment which have been identified above. Consequently, at the level of immediate experience (prior to any reflective objectification of body) illness manifests itself essentially as a disruption of lived body.

Bodily dysfunction necessarily causes a disturbance in the various and varying interactions between embodied consciousness and world. Thus, the very nature of body as being-in-the-world is transformed.[88] First and foremost illness represents dis-ability, the "inability to" engage the world in habitual ways. A headache is not experienced simply as a pain in the head, but as the "inability to" concentrate on the book I am reading, enjoy the music I am listening to, have an animated conversation with my spouse, and so forth. Arthritis represents not so much an inflammation of the joints as it does the "inability to" button my shirt, swing a golf club, play tennis. In the event that the illness is chronic or life-threatening this experience of dis-ability relates not only to one's immediate engagement in the world but portends the "inability to" carry out future projects or to complete anticipated goals.

In illness bodily intentionality is frustrated. Objects which were formerly grasped as utilizable (and were thus largely taken-for-granted and unnoticed) now present themselves as problems to the body. For the person with angina, for example, a flight of stairs which in health was simply there "to be climbed," is now perceived as an obstacle "to be circumvented," "avoided," or even "feared." Habitual acts (such as walking, running, lifting, sitting up, eating, talking, and so forth), which were hitherto performed unthinkingly, now become effortful and must be attended to.[89] Thus, the sphere of bodily action and practical possibility becomes circumscribed. The "I can" is rendered circumspect.

The relation between the lived body and the environment is also changed in that, with illness, the surrounding world looks and feels different. If one has a migraine headache, for example, one experiences the sunlight streaming through the window as intolerably bright where hitherto it had seemed warm and inviting. When sick, one may be nauseated (rather than tempted) by cooking odors; indifferent to, or repulsed by, the sight of formerly mouthwatering displays of food. Sounds are unusually harsh or, alternatively, distantly muted outside the sickroom. The touch of material against the skin or the weight of the bedclothes on top of the body may be experienced as unpleasant or even painful. In illness the world impinges upon the senses in unfamiliar ways rendering uncomfortable one's being as a being-in-the-world.[90]

Body image changes, not only in terms of such things as posture, gait, and so forth, but in the sense that one no longer has available "an open system of an infinite number of equivalent positions directed to other ends." The possibilities for action shrink. If I am ill I simply do not have available to me all the alternatives that are available in health. Whether I like it or not, there are certain activities, postures, gestures, and so forth, which are no longer within my bodily scope.

Additionally, the primary meaning provided by the body may be disrupted. The multiple sclerosis patient who trips on the stair and the visually impaired person who walks into the table, for example, both find their body's intuitive sense ineffective, indeed deceptive. The primitive spatiality of the body has been disturbed. The body no longer correctly interprets itself and the world around it. In this event the physiognomy of the world has changed.

Moreover, in response to the demands of illness, the contextural organization of the body and the complexure of body/mind/world must shift in varying ways. Such adjustments in the contexture are experienced as "foreign," alien, unnatural. "I'm just not myself today," or "Things just don't feel right," express in part this perceived change in bodily experience. Also, the figure/ground relation of body/world changes in illness. As Mary Rawlinson (1982, p. 75) notes, "whereas our embodied capacities ordinarily provide the background to the figure of our worldly involvements, in illness our body, and particularly that aspect which pains, becomes itself the figure of our intention against which all else is merely background."

Illness also effects a change in the body's gestural display. This change may be perceived by others in the "look" of the sick person, the "grimace" of pain, the stoop of the shoulders, the limp. The one who is ill is uncomfortably aware of the change in gestural display. After all, as Merleau-Ponty has pointed out, we develop a certain corporeal "style" which identifies the body as peculiarly "mine." A limp may appear scarcely noticeable to an onlooker, yet be profoundly disturbing to the one who develops the limp. It represents a fundamental change in his or her body style.[91] In addition, such gestural displays as weeping, laughter, grimacing, and so forth, may be experienced as uncontrollable and inappropriate by the patient and may be misunderstood and shunned by others.[92] In this respect, it is important to note that not only do bodily patterns of walking, talking, gesturing, and so forth, identify the lived body as peculiarly "mine" but such patterns represent my self-presentation to others.[93] Consequently, it is particularly hard for patients to accept changes in body style which project a negative body image (i.e., a body image which does not match up to cultural ideals of attractiveness, physical fitness, and so forth).[94]

Gestural display is an integral element in the many forms of social interaction and expression which we take for granted – standing facing one another, shaking hands in greeting, gesticulating, and so forth. Disruption of these usual modes of communication results in a concurrent disruption of the social world. A transformation in body style is not only discomforting to the afflicted person but it is a source of discomfort to others. The stroke victim with a paralytic arm finds others embarrassed at the loss of function. How does one shake hands in greeting? Eyes are typically averted from the person with a severe tremor or disordered body style. Consequently, an alteration in the body's gestural display can make the patient's relations with others awkward where, hitherto, social interaction had been effortless.

Zaner suggests that upright posture itself has significance as a gestural display. He notes that Straus' analysis of the meaning of lateral space shows upright posture to be "pregnant with a meaning not exhausted by the physiological tasks of meeting the forces of gravity and maintaining equilibrium" (Zaner, 1981, p. 61). As Straus (1966) has shown upright posture is the crucial element in the constitution of lived spatiality and in the possible modes of communication and performance.

The value assigned to upright posture should not be underestimated when considering the experience of illness. To be able to "stand on one's own two feet" is of more than figurative significance. Verticality is directly related to autonomy. Just as the infant's sense of autonomy and independence are enhanced by the development of the ability to maintain an upright posture and "sally forth" into the world unaided, so there is a corresponding loss of autonomy which accompanies the loss of uprightness. Indeed, not only does loss of verticality (upright posture) engender feelings of helplessness and dependency in the one who is ill, but it causes others to assign the dependent role to the patient. One has only to spend a morning in a wheelchair, or a day in a hospital bed, to experience firsthand the loss of integrity and autonomy which accompanies the loss of upright posture. As an M.S. patient, for example, I am intrigued by the fact that, when I am in my wheelchair, strangers tend to address themselves to my husband and refer to me in the third person. "Would *she* like to sit at this table?." "What would *she* like to drink?." and so forth.[95]

There is more than metaphorical significance to such expressions as "to look down on" and "to look up to." In the hospital setting the patient, more often than not, is in bed and must "look up to" the doctor who "stands" talking and "looking down on" the patient.[96] In "looking up to" the doctor, and "being looked down on," the patient feels on an unequal "footing" with the physician, concretely diminished in autonomy.[97] As the physician, Edward Rosenbaum, commented on his own hospitalization:

Lying in bed in a hospital room was a new experience.... I had been in similar rooms thousands of times, but in a different position. Then I was in command, neatly dressed, standing, looking down at a helpless patient in bed. Now I was that patient, literally stripped of my dignity. I was no longer in charge. I was being treated like a baby (Rosenbaum, 1988, p. 5.)

In this regard it is worth noting that patients are likely to feel much less "inferior" if the physician sits down by the bedside, so that they are on the same level ("eye to eye") when communicating with one another. Robert Kravetz, a practicing gastroenterologist writing of his own illness, cites (1987, p. 434) a study that compared patients' perceptions of how long their doctors spent with them when they stood or sat by the bedside. Although the time spent in the study was exactly the same, the patient always perceived that more time was spent when the physician sat. The conclusion drawn from the study was that because it was perceived that the physicians had spent more time when sitting, it was also felt that they were more interested and concerned about the welfare of their patients. Speaking of his personal experience, Kravetz says:

I noticed that those visitors who sat at my bedside and chatted with me seemed to be spending more time with me and I felt that they were more interested in my well-being. I had always made it a practice to sit with my patients at the bedside, and after being in this position myself, I heartily endorse this type of patient visit because it creates a much more intimate physician-patient relationship and one of caring concern (Kravetz, 1987, p. 434).

I would argue that it is not simply a matter of creating a more intimate and caring relationship between patient and health care provider. Rather, it is important to recognize the profound effects of the loss of upright posture and to take steps to mitigate these effects.[98]

In illness the character of lived spatiality changes in significant ways. As the foregoing description of lived body reveals, the spatiality of the body is not a spatiality of physical location but a spatiality of situation. If I place my arm on the table, for example, I do not think of the arm as being "beside" the ashtray in the same way that I consider the mug to be "beside" the ashtray. Rather, my body appears to me as "an attitude directed towards a certain existing or possible task" (Merleau-Ponty, 1962, pp. 98–100). I place my arm on the table "in order to" – in order to reach for the mug, to put a cigarette in the ashtray, and so forth.

Physical space is thus for my body an oriented space. The objects which surround me necessarily refer back to my bodily placement, my orientation, within the world. The placement of objects is not defined simply by purely spatial coordinates but rather is defined in relation to axes of practical reference. "The glass is on the coffee table" means I must be careful not to upset it if I move the table. "The chair is to the right of the desk" means that I must avoid bumping into it when I walk past the desk to go to the door (Sartre, 1956a, p. 424).

Physical space is thus presented to me as functional space, as that milieu within which I am able to perform my various activities. Points in space do not represent merely objective positions in relation to the objective location of my body. Rather they mark the varying range of my aims and gestures. For example, the narrow doorway through which I must pass presents itself to me not as an object but as a "restrictive potentiality" for my body requiring modification of my actions (Merleau-Ponty, 1962, p. 143). My embodying organism is always experienced as "in the midst of environing things, in this or that situation of action, positioned and positioning relative to some task at hand" (Zaner, 1981, p. 97).

In the experience of illness the character of lived spatiality changes. In the normal course of events locomotion continually opens up space, allowing one freely to change position and move towards objects in the world. Illness and debility exert a centripetal force anchoring one in the Here. If I am in bed with the stomach 'flu or confined to a hospital room following surgery, I experience a concrete shrinking of my world (to the confines of my bed, my room, my house, and so forth). This restriction to the Here also engenders a heightened sense of distance between oneself and surrounding things. In one's changed physical condition, one finds that objects or locations which were formerly regarded as "near" are now experienced as "far." For instance, the bathroom which appeared "near" to the bed in health is suddenly experienced as "far" from it when one is ill. Friends and colleagues recede into the distance. The workplace seems a "world" away.

Spatiality also constricts in the sense that the range of possible actions becomes severely circumscribed. Rather than representing the arena of possible action, space is encountered as the restriction of possibilities. Projects may have to be modified or perhaps set aside altogether. The "in order to" character of bodily gearing into the world must be explicitly attended to, often with unaccustomed effort. In this regard, *functional* space assumes an unusually problematic nature. Ordinary objects are encountered as "restrictive potentialities" for the body. For example, for the person with a tremor the mug is no longer simply there "to be grasped" but, rather, presents itself as a problem to be solved. Stairs are obstacles not only to the person with angina but to the person with M.S, the man on crutches, and the patient recovering from a severe bout of influenza. For the visually impaired, surrounding objects represent constant reminders of the inability to focus, while crossing a busy street becomes a nightmare. For the person with a hearing loss a telephone conversation or a social gathering is an ordeal. The examples can be continued *ad infinitum*. The crucial point is that in varying ways the ill person perceives a distinct change in the spatiality of the body (the spatiality of situation) – a change which permeates the surrounding world. Furthermore, since familiar objects such as stairs, doors, tools represent concrete experiences of bodily limitation and the frustration of intentionality, illness renders explicit one's being as a being-in-the-world. A problem with the lived body is a problem with the environment because world and body represent a unified system.

Physical space itself assumes a restrictive character. Slopes may be too steep to climb, sidewalks too uneven to walk on, doorways too narrow to navigate with a wheelchair. In illnesses which involve an ongoing loss of function, the restrictive character of physical space is a permanent feature of the patient's lived experience. Consequently, people with disabilities necessarily view the world through the distorted medium of their damaged bodies.[99] As a person with limited mobility, for instance, I can remember that my first impression of the Lincoln Memorial was not one of awe at its architectural beauty but rather dismay at the number of steps to be climbed.[100]

Illness causes a disruption not only in the character of lived spatiality but additionally results in a change in temporal experience.[101] Just as lived spatiality is characterized by an outward directedness, a purposiveness and intention, so time is experienced not as a static present but as a moving towards the future. Normally we act in the present in light of more or less specific goals which relate to future possibilities. In illness such goals suddenly appear irrelevant or out of reach. One finds oneself preoccupied with the demands of the here and now, confined to the present moment, unable effectively to project into the future. Indeed, life projects may have to be set aside, modified or abandoned altogether. As Zaner (1981, p. 176) notes, illness obstructs the human ability to "possibilize," to free oneself from the actual in order to move to the possibly-otherwise.

In this respect the effortful nature of everyday existence engendered by illness has an impact on the subjective experience of time. In speaking of his progressive blindness, for instance, John Hull (1990, pp. 78–80) makes the interesting point that illness and dis-ability result in a changed relation to time. Rather than "trying to cram every minute with necessary tasks and to squeeze the last drop out of time" in order to get to the next moment, one is forced to concentrate on the present moment and the present task. "You are no longer fighting against the clock but against the task. You no longer think of the time it takes. You only think of what you have to do. It cannot be done any faster" (Hull, p. 80). In forcing one to concentrate on the actual, illness prevents one from projecting oneself into the future moment. "The reason why I do not seem to be in a hurry," says Hull, "is not that I have less to do than my colleagues, but I am simply unable to hurry."

This is not just the case with blindness, of course.

I think of my friend Chris with his multiple sclerosis.... Time ... has strangely expanded. It takes him forty-five minutes to tie up his shoelaces in the morning. It doesn't matter. He does not get impatient. He just does it. That is how long it takes to tie shoelaces. I think of Clive Inman, with his back injuries.... Space to him is diminished to the size of his bed. On the other hand, for those twelve *long* weeks, he has all the time in the world (Hull, 1990, pp. 79–80). (Emphasis mine).[102]

The significance of past, present and future may change in other ways in illness. For example, a dire prognosis for the future may be perceived as an imminent and ever-present threat. The uncertainty associated with progressively degenerative diseases such as multiple sclerosis, for example, may cause the newly diagnosed patient to start living as if *already* severely incapacitated, or as if the threat is immediate. In this case temporality is disrupted in a different way. It is not simply that there is a loss of the future through the loss of future goals, but that the illness engenders a concurrent loss of the actual present. The actual present is forfeited and transposed into an imagined future.

In the case of life-threatening disorders, the future disappears. After being diagnosed with cancer, Frank (1991, p. 37) noted that all he could see were the faces he would never grow old with – his daughter, his wife, his parents. "The pain of my death," he says, "was in losing my future with those others."

Similarly, the significance of the past may take on a different character in illness. A remembered past event which was very threatening, such as a severe attack of illness, surgery, or an accident, may be perceived as close in time. "It can't be two years since I had my heart attack." This past

threat may also pervade the present. The patient may live in constant fear
of a recurrence – a fear which may increase instead of diminishing as time
passes. "It's two years since I had an attack of my illness. It's about time
for another."[103] As Cassell (1985, p. 28) notes, the meaning of the
objective time scale is subjective and varies with the patient (and, I would
argue, with the type and stage of illness).

## *Summary*

In summary, then, at the pre-reflective level illness is experienced as a
disruption of *lived body*. The fundamental features of embodiment, such
as being-in-the-world, bodily intentionality, primary meaning, contextural
organization, body image, and gestural display, are all disturbed in
various ways. The character of lived spatiality and lived temporality
undergoes a significant change. Consequently, illness-as-lived represents
a chaotic disturbance and sense of disorder in the patient's being-in-the-
world. Furthermore, since at the level of lived body there is no perceived
separation between body and self (at this level I *AM* my body), illness
necessarily incorporates not only a threat to the body but a threat to one's
very self.

## 4. BODY AS OBJECT IN ILLNESS

As has been noted, under normal circumstances the body appears as an
object both in the experience of being an object for another and in certain
"limit situations" in which the body is apprehended as a material, physical
entity. Such bodily objectification separates self from body and, depending upon the circumstances, may result in a deep sense of alienation from
one's body. In examining the manner in which the patient apprehends the
body-as-object, it will be noted that the objectification of body is an
integral element in illness. Such objectification is necessarily accompanied by feelings of both alienation from, and unwilling identification
with, the body.

Illness represents a "limit situation" in which the body is apprehended
both as a material, physical entity and as a being-for-the-Other. In the first
place, illness engenders a shift of attention. The disruption of lived body
causes the patient explicitly to attend to his or her body *as* body, rather

than simply living it unreflectively. The body is thus transformed from lived body to object-body.[104] This objectification results in the apprehension of the corporeal nature of the body as a physical encumbrance, as an oppositional force, as a machine-like entity and as a physiological organism.

For example, in the normal course of events when reaching for my cup of coffee to take a drink, I do not explicitly focus on the action of my hand. Rather, my attention is directed to the task at hand (lifting the cup). However, should I injure my hand, then my attention is focused on my hand *as* hand. I must observe how it is that my fingers grasp the handle of the cup and I am conscious of my hand's unaccustomed ineffectiveness as an instrument of my actions. In illness the body intrudes itself into lived experience. It becomes the focus and object of scrutiny. Furthermore, with the breakdown of function, the instrumentality of the body announces itself. For example, if I cannot see properly I perceive my eye explicitly as an instrument-for-seeing and, more particularly, as a *defective* instrument-for-seeing.[105]

In apprehending the body explicitly as an "instrument-for" actions within the world, the patient perceives it to be a material, "physico-biological thing." Furthermore, with dysfunction it is perceived as a defective "physico-biological thing." Therefore, the patient objectifies the body not only as a physiological organism but as a *malfunctioning* physiological organism. As is the case with "disease," this apprehension of the object-body will reflect the particular lifeworld in which the patient is situated (and indeed the stage of illness). For example, initially the body-as-object may simply be conceived in terms of faulty mechanism (i.e., there is an apprehension that the machine-like, physical body simply doesn't "work right"). The patient's conception of the dysfunctional body will, however, reflect the theoretical understandings that are embedded in a particular lifeworld.[106] If, for example, I have blurred vision I will recognize my eye not only as a defective "instrument-for-seeing" but I will incorporate into my understanding of the eye's "not working right" some conception (albeit sketchy and incomplete) of the anatomy and physiology of vision. If I have chest pain and I have a history of heart disease, I may incorporate into my conception of physiological body some explicit reference to narrowed coronary arteries. This understanding necessarily reflects my particular cultural background and the meanings inherent in my unique biographical situation.

The malfunctioning body is further apprehended in terms of its mechanistic nature in that it is perceived by the patient to be a machine-like entity comprised of organ systems and parts, some of which can be repaired, removed or technologically supplemented.[107] I can, for example, rid myself of an inflamed appendix or a cancerous breast. Under extreme conditions I can even have my heart cut out to be replaced by a transplanted organ. Obviously, not all parts of this complicated mechanism are expendable or restorable and, to the extent that this is the case, I cannot disassociate myself from my machine-like body in toto[108] – nor can I detach myself from certain parts of it (such as the central nervous system).[109]

In addition, the corporeal nature of the malfunctioning body is rendered explicit in that the body becomes an oppositional force in illness. One may, for example, concretely experience the heaviness of one's limbs, the resistance of stiffened joints, the powerlessness of weakened muscles, the contrariety of trembling hands. The sheer physicality of the body impedes one's interaction with the world providing inert and overt resistance. Rather than being that which enables one carry out one's intentions in the world, the physical, material body presents itself as an impediment which must be overcome. In this experience the object-body may be apprehended without explicit reference to pathoanatomical concepts. Rather, it is a direct apprehension of physical encumbrance.

The experience of the body as a physical encumbrance is obviously most evident in illnesses which involve the overt loss of function. Illnesses which manifest themselves in terms of a change in appearance, such as an unusual rash or a lump in the breast, present the body as oppositional (in that the malfunctioning body opposes the self in disrupting one's ongoing plans and projects) but in these circumstances the opposition is less likely to appear as a pure physical resistance.

The objectification of the body not only as physical encumbrance but, more particularly, as a *malfunctioning* physiological organism further contributes to the sense of bodily alienation which characterizes illness. In particular, this renders explicit the experience of the body as "uncanny." Bodily dysfunctions disclose the latent implications of embodiment, and reveal what it means to be embodied. While the sense of "otherness" of body is by no means peculiar to illness, it is concretely felt in this experience.

It will be recalled that Zaner identifies four senses in which the body is experienced as "uncanny" under normal circumstances (corporeal implicatedness, inescapable/limitation, hidden presence, and alien presence). I would argue that the foregoing analysis reveals that these four senses of the "uncanny" are concretely realized in the experience of illness, and particularly in the experience of the body as a malfunctioning physiological organism.

As we have noted, as a malfunctioning physiological organism, the body is out of one's control in important ways. This lack of control reveals the symbiotic relation between body and self. In illness one comes face-to-face with one's inherent vulnerability and dependence upon one's body. "It could really happen to me!" is felt as a concrete actuality and not simply as an amorphous possibility. Moreover, try as one might, one cannot altogether disassociate oneself from the malfunctioning body. One must explicitly take it into account as a precondition of one's plans and projects. Thus, one is uncomfortably aware of the contingency of being embodied in this particular malfunctioning body and the implications of such embodiment in terms of the threat to the self. The sense of inescapability and limitation are intrinsic to illness-as-lived.[110]

Moreover, the experience of the body as a malfunctioning physiological organism overtly discloses the body as an entity which includes physiological processes, events and structures over which I have limited control. Indeed, in many cases, I recognize that I am not (nor can I be) even aware of such processes and structures. Thus, in illness (more particularly than in health) one explicitly experiences the "hidden" and "alien" presence of the body.[111]

It is interesting to note that patients often verbally express their awareness of the body as alien presence. Cassell notes, for example, that patients will attempt to disassociate themselves from their bodies in describing their symptoms. Rather than referring to "my" leg, or "my" breast, they will refer to "the" leg or "the" breast. "If you see something on the mammogram does that mean 'the' breast has to be removed?" Cassell (1985b, pp. 55–65) suggests that the use of impersonal pronouns is a means to avoid contact with an intrusive reality. I would suspect that it also represents the patient's feelings of alienation with respect to the body.[112]

In this regard it should perhaps be pointed out that the constitution of body-as-other may in some circumstances be desirable. It may, as Cassell implies, be less traumatic to consider removal of "the" breast, rather than removal of "my" breast. Furthermore, in regarding my body as a mechanistic object I can (paradoxically) in some situations regain some control over it. I can, for example, have an obstructed artery cleared, a broken arm repaired, a malfunctioning heart valve replaced. Nevertheless, it should be noted that this sense of control is tenuous. The cleared artery may become obstructed again in the future or the broken arm may fail to heal, regardless of my (or my physician's) best endeavors. Consequently, the constitution of body-as-other (even in instances where such constitution is desirable) at some level involves the apprehension of body as alien presence.[113]

The shift of attention which renders the body thematic in illness is necessarily a part of the clinical encounter. In order to cooperate with the physician the patient must explicitly attend to his or her body as object (in the giving of an "objective" report of body sensations, in self administering treatments and reporting back on all changes in the external appearance and internal sensations of the body, and so forth) (Leder, 1984a, p. 33). The lived body is thereby transformed. For example, as Sartre (1956a, p. 403) notes, in observing my leg as an object what I cause to exist is the *thing* "leg"; it is not the leg as the *possibility which I am* of walking, running or of playing football.

Moreover, in the clinical encounter the body is objectified not only as a material, physical entity but as a being-for-the-Other. Under the "gaze" of the physician, the patient perceives his or her body to be an object of scientific investigation. In the experience of being looked-at by the physician, one recognizes not only one's being-an-object for the Other but the brute fact of one's being as a biological entity. To undergo the experience of being taken as an "object" by the Other is to experience concretely the "ambiguity" of own-body (i.e., to experience the strange duality of being at once subject for oneself and object for the Other).[114] As has been noted, such experience is in no way limited to the clinical encounter. However, what perhaps sets the clinical encounter apart is the fact that in the course of a medical examination the patient experiences not only his or her being as an object for the Other but, more specifically, his or her being as an object of scientific investigation. Consequently, the patient finds not only body but self reduced to a malfunctioning biological organism. Furthermore, in discussing the illness with the physician, the patient is acutely aware that there is a disparity between one's experiencing as a subject and one's being experienced as an object.

Sartre has noted that at the level of "disease" the patient apprehends *illness* as a being-for-the-Other, in that "disease" is known to the sick person by means of concepts derived from others (such as the principles of physiology and pathology described by others). To the extent that it incorporates some reference to pathoanatomically based theoretical constructs, the apprehension of the *body* as a malfunctioning neurophysiological organism likewise represents a being-for-the-Other, in that it too involves bits of knowledge acquired from others.

## Summary

In sum, then, it is clear that the objectification of body is an integral element in the lived experience of illness. The body is objectified by the patient both as a malfunctioning physiological organism and as a physical encumbrance. As a malfunctioning physiological organism the damaged body appears as a defective instrument or a faulty tool (which is clearly not the way it appears to us in health even in those moments when we are aware of the physical nature of our bodies). In addition, in the clinical encounter the patient construes the body-as-object in the concrete experience of being-for-the-Other.

What is peculiar about bodily objectification in illness is that the apprehension of body-as-object is such that it renders the experience of "uncanniness" explicit, often resulting in a profound sense of alienation from body. This is not necessarily the case in normal circumstances where the otherness of body may be a positive experience. Moreover, to the extent that one is forced to take the impaired body into account in carrying out one's projects in the world, so the experience of alienation and "uncanniness" is ever-present. This experience is a paradoxical one. My body appears as Other-than-me in that it opposes and frustrates my intentions; yet I *am* my body for I cannot escape my impaired embodiment.

While the paradoxical relation between body and self is explicitly recognized in all forms of illness, it is felt most profoundly in chronic illness. As we have seen, the objectification of body in illness results from a forced attention to physical function and the awareness of some impairment or other physical change. In chronic illness this forced attention to body is a daily occurrence. As a multiple sclerosis patient, for example, even though I have adapted to my physical disabilities, I must overtly take them into account as I go my way about the world. On a daily basis, whether I like it or not, I am aware of my dysfunctional body as both physical encumbrance and as malfunctioning physiological organism. This is obviously the case for all those who suffer from chronic ailments which disrupt everyday functioning on a regular basis.

The prolonged attention to body which occurs in chronic illness engenders a kind of metamorphosis. The body is transformed into a new entity, the "diseased body."[115] The "diseased body" with its ongoing demands, necessarily stands in opposition to the self. One must continually compensate for its dis-abilities, allow for its weaknesses, pay unwilling attention to its pains, and so forth, before one can carry out one's projects in the world.

Indeed, the bodily metamorphosis which occurs in chronic illness incorporates a sort of Gestalt switch in that the experience of bodily disruption becomes one's normal expectation and non-disruptive moments appear as somewhat fleeting anomalies. In this sense the "diseased body" (rather than being that which is routinely overlooked in carrying out one's projects in the world) is experienced *on an ongoing basis* as an insistent presence against which all else is background.

Not only is the "diseased body" constituted as a malfunctioning physiological organism but with chronic illness there can be no expectation of a return to normal function. One perceives one's body to be permanently impaired. Consequently for the chronically ill the sense of alienation from, and unwilling identification with, body is particularly profound.

## 5. THE BODY-AS-SCIENTIFIC-OBJECT

In considering the manner in which the body is intended differently by physician and patient, it is helpful to recapitulate briefly the analysis of illness in the foregoing chapter. It will be recalled that, in apprehending illness as a disease state, the physician (as natural scientist) thematizes the patient's immediate experience of bodily disruption in terms of theoretical, scientific constructs such that the experience is wholly subsumed under the causal categories of natural scientific explanation. Symptoms thereby become re-interpreted as physical signs (visible lesions) and physiological processes are translated into objective, quantified data (lab values, images, graphs, numbers, and so forth). Since disease is categorized in the same way as other natural phenomena, it can be viewed independently from the person suffering from the disease.

It will also be recalled that the disease state, as construed by the physician, is not identical with the "disease" which is apprehended by the patient. Although the patient may come to view his or her "suffered illness" as "disease" (a view which may incorporate some reference to pathoanatomically based theoretical constructs and which involves some assignment of explanatory meaning to the immediate experience of bodily disruption), "disease" is still an amorphous entity which is not directly experienceable. For example, as was noted, although as an M.S. patient I may come to recognize the numbness in my arm *as* "multiple sclerosis" and, further as involving some disruption of certain sensory pathways, I do not directly experience the lesion in the central nervous system which is the disease state known by the physician. For the patient the fundamental entity of illness is the body painfully-lived whereas for the physician the fundamental entity is the disease state.

A similar distinction exists in the apprehension of the patient's body by physician and patient. Under the medically trained "gaze" of the physician, the patient's lived body assumes the status of a scientific object (i.e., it is intended as a neurophysiological organism and, more particularly, as a mass of cells, tissues, organs, and so forth according to the categories of natural science.) The human body which is presented to the physician in the clinical encounter is understood by the doctor to be a strictly biological affair, ultimately explainable in purely physical terms. The medical eye focuses on the various bodily systems, organs, structures, and functions in an effort to render explicit the inner workings of this complicated neurophysiological mechanism – and thereby to pinpoint the disease state.

It is important to note several things about the physician's conception of the body. In the first place the body-as-scientific-object is thematized within the naturalistic attitude. Thus it is conceived as a purely physical, material thing whose mechanism is wholly explicable (or in principle explicable) in terms of the categories of natural science. Thus, this particular body presented to the physician in the clinical encounter is simply an exemplar of *"the"* human body and, as such, it may be viewed independently from the person whose body it is. That is, the mechanical workings of this particular human body are "objectified" in such a way as to render the "subjective" experience of the particular patient explicable in terms of a general, theoretical account of the causal structure of such experiencing. As Husserl (1970a, pp. 315–383) has noted, the goal of natural science (and the purpose of the scientific attitude) is not to live in reality as it is experienced by particular persons but rather to explicate reality in terms of universal, causal laws which are "objective" and thereby valid for all.

Foucault (1975, p. 111) and Zaner (1988, pp. 154–170) have noted that according to the modern, scientific understanding of disease, under the "gaze" of the physician the body-as-scientific-object is transformed from lived body to anatomical body and, as such, it assumes the guise of a corpse. It will be recalled that, largely as a result of developments in the science of pathoanatomy in the 19th century, the primary focus of medicine went inside the body and disease thus became identified with pathoanatomical lesions or pathophysiological disturbances (Engelhardt, 1982, pp. 41–57). The live body thus became explicable in terms of the dead body.

Now for the first time, indeed, it became possible for medicine to offer what was seen as the genuinely scientific explanation of what is strictly individual; anatomical experience at correlating tissual lesions with formerly observed (and recorded) clinical symptoms permitted increasingly controlled inferences from the latter to the former. And, with this, clinical observation itself became increasingly a matter of "autopsy-in-advance," with observations becoming anticipations of what actual autopsy would eventually find (Zaner, 1988, p. 134).

In particular, it is important to note that the body-as-scientific-object is no longer the totality "body in situation" as is the case with the lived body.[116] The body of anatomical-physiology does not represent the synthetic unity of a particular life, the embodiment of a unique individual. Rather than representing a being-in-the-world, the physical body (as a scientific object) is taken purely in terms of its mechanistic nature and is, thereby, no longer "in situation."

In apprehending the body-as-scientific-object, the physician is concerned to move beyond the outward appearance of the material body to detect its innermost workings. In other words the "medical gaze" is directed to the inside of the body.[117] In the physical examination the physician interprets the surface signs of the body (edema, cyanosis, and so forth) as merely the outward manifestation of the pathology within. With the aid of various technologies (stethoscope, ophthalmoscope, and so forth) and through the use of his or her own perceptions (e.g., the probing fingers which feel the abdominal mass, the eye which sees the unsightly rash, the ear which hears the heart murmur) the physician turns attention from the surface to the body's interior. This process may be further assisted through the use of machines (x-rays, CT scans, and so forth) which actually visualize organs and structures located deep within the body. Furthermore, in diagnostic workups physiological processes may be reduced to lab values, numbers, graphs and so forth in an effort to render such processes explicit. In regarding the body as a scientific object, then, the physician in a sense renders the outer appearance of the physical object-body transparent. Under the physician's medically trained "gaze" the object-body is conceived in terms of its interior (i.e., the body-as-scientific-object is conceptualized as a mass of cells, tissues, organs which comprise the material body).

In discussing the separate worlds of physician and patient, I noted that the manner in which an object is thematized is directly correlative to the way in which an individual attends to that object (such attentional focus being determined within a particular context of relevance).[118] I further noted that within the context of the scientific attitude and in light of medical training, the physician construes the "reality" of the body in a manner which reflects the "habits of mind" of the medical profession. Consequently, just as a painting is viewed differently by a professional artist as opposed to the "man-in-the-street," so the body is "seen" differently by the experienced physician. For example, the cardiologist hears the heart murmur, feels the substernal thrill that is undetected (and undetectable) in the non-medical "gaze" (Leder, 1990a). This manner of thematizing the body in the scientific attitude is quite distinct and represents what Schutz has called an "autonomous province of knowledge."

It is important to emphasize that the body as intended within the scientific attitude by the physician is significantly different from either the lived body or the object-body intended by the patient. In the first place, as has been noted, the body-as-scientific-object is wholly subsumed under the categories of natural scientific explanation (i.e., it is apprehended with explicit and exclusive reference to pathoanatomically based theoretical constructs). As such, the anatomical body represents not the lived body (one's intentional being and mode of access to the world) but rather the cadaver which may be dissected at autopsy. Furthermore, as a scientific object, a particular body is simply an exemplar of *the* human body (or of a particular class of human bodies) and, as such, it may be viewed independently from the person whose body it is.

For the patient, however, the body does not represent a scientific object. Indeed, at the pre-reflective level one is not aware of one's body *as* body (i.e., one does not objectify it as a neurophysiological organism nor pay attention to its mechanistic nature). At the level of lived body I simply "exist" my body. The lived body represents my being-in-the-world and illness is fundamentally experienced as a disruption of this embodiment.

At the reflective level the objectification of body by the patient reveals the material, physical nature of body and particularly the instrumentality of the body. As a defective instrument, or faulty tool, the body may appear not only as a material, physico-biological thing but as a malfunctioning physiological organism. Nevertheless, this apprehension by the patient is not identical with the physician's conception of body-as-scientific-object. Rather, the patient's apprehension represents the experience of the "uncanny," of one's own body as hidden and alien presence. Although this may include some reference to pathoanatomically based theoretical constructs (some reference to the objective nature of "the" body), the patient recognizes this body to be "his" or "her" body and thus the patient cannot totally disassociate from it. Thus, the objectification of body as malfunctioning physiological organism by the patient involves explicit reference to the body as "owned" and incorporates a sense of both unwilling identification with and alienation from one's body.[119] Furthermore, just as one does not directly experience the "disease state," so one does not directly experience the body-as-scientific-object (the body known by the physician).

The apprehension of the body as "uncanny" is also evident when the patient is a physician. For instance, Fitzhugh Mullan (1975, p. 4) notes that on viewing the chest x-ray which revealed his own cancer he "instinctively looked at the grim information on the viewing box as a clinician" but in a matter of minutes he began to:

[C]ome to grips with what was happening. That pint-sized cauliflower that I had so recently discovered on a piece of celluloid was in fact a tumor – a cancer. It was living quietly deep within MY body ... in a space of five minutes it had come out of nowhere to become the focal point of my life, or perhaps the focal point of the rest of my life (Mullan, 1975, p. 4).

Other autobiographical accounts by physicians make it clear that although they have the training to turn the "medical gaze" upon their own bodies, the experience of illness is such that the bodily objectification of one's own body is fundamentally the experience of the body as "uncanny" (Mandell and Spiro, 1987).

For the physician *qua* physician, then, the fundamental entity is the body-as-scientific-object. As such the patient's body represents simply an exemplar of "the" human body (and indeed of the human cadaver) which may be explicated wholly in terms of the concepts of natural science. For the patient, however, the fundamental entity is the body "painfully-lived." The body "painfully-lived" represents not only the immediate experience of bodily disruption at the pre-reflective level but the apprehension of "uncanniness" at the reflective level.

### 6. IMPLICATIONS FOR MEDICAL PRACTICE

The foregoing analysis of the manner in which the body is intended has some important implications for medical practice. In the first place it is evident that illness is fundamentally experienced by the patient as a disruption of lived body. Consequently, illness must be understood not simply as the physical dysfunction of the mechanistic, biological body but as the disorder of body, self and world (of one's being-in-the-world). Unlike the conception of body-as-scientific-object, the paradigm of lived body situates illness in the particular patient in a very explicit way. The biological body cannot be conceived as separate from the person whose body it is. The biological body represents this patient's particular embodiment and, as such, this embodiment bears certain relations to a particular world and to a unique self. The patient does not simply "possess" this body. He or she *IS* this body. Consequently, patients do not so much "have" a bodily illness as they "exist" their illnesses.[120] For example, people who live with multiple sclerosis, arthritis, heart disease and so forth, are persons living a disordered existence in very specific ways, not just persons who "have" certain identifiable diseases. As lived, the body must be conceived as body-in-situation. A dysfunction in biological body represents a concurrent disruption of the patient's being-in-the-world. "What happens when my body breaks down happens not just to that body but also to my life, which is lived in that body," says Arthur Frank (1990, p. 8). "When the body breaks down, so does the life."

Indeed, in his books Sacks (1983; 1985c) has shown how illnesses may be understood in terms of the "organized chaoses" which they produce in the patients' worlds. In presenting clinical studies of various neurological disorders he has grouped such disorders not according to the traditional classifications of neurology but in terms of such disturbances of world as "losses," "excesses," and "transports" (Sacks, 1985c). Such clinical studies not only provide insight into the lived experience of illness but they suggest therapeutic approaches to the disorders. Moreover, as Sacks notes:

[A] disease is never a mere loss or excess ... there is always a reaction, on the part of the affected organism or individual, to restore, to replace, to compensate for and to preserve its identity, however strange the means may be: and to study or influence these means, no less than the primary insult to the nervous system, is an essential part of our role as physicians (Sacks, 1985c, p. 4.)

The patient comes to the physician because of a perceived disruption in everyday life, a sense of disorder of embodiment. As Edmund Pellegrino and David Thomasma (1981, p. 72) show, the goal of medicine is primarily the relief of this perceived lived body disruption – the restoration to a former or better state of perceived health or well-being. This may include, but is not limited to, cure of organic dysfunction. Indeed, in order to address the patient's experience of disorder, attention must be paid not only to the physical manifestation of a disease state but also to the changing relations between body, self and world. This is, perhaps, especially the case in chronic illness where the disintegration of self and world is felt most profoundly.

The recognition that there are certain essential features to embodiment (such as being-in-the-world, bodily intentionality, contextural organization, body image, gestural display, and so forth) provides a clue as to the manner in which illness manifests itself as a disorder of embodiment. In particular, direct attention can be paid to such disturbances as the change in lived spatiality and lived temporality.

As was noted, illness causes a constriction in the lived spatiality of the patient in that the range of possible actions becomes severely circumscribed and physical space itself takes on a restrictive character. In recognizing that physical space represents functional space, Merleau-Ponty (1962, p. 145) notes that in the normal course of events (through the performance of various habitual tasks) the embodied individual incorporates objects into bodily space. For example, the woman who habitually wears a hat with a long feather intuitively allows for the extension of the feather when she goes through a doorway. The experienced typist no longer views the keys of the typewriter as objective locations at which she must aim. Rather, the person who knows how to type incorporates the key-bank space into his or her bodily space. Similarly, the blind man's stick after a time ceases to be simply an object but becomes an extension of body increasing its range. The point of the stick becomes "an area of sensitivity, extending the scope and active radius of touch, and providing a parallel to sight" (Merleau-Ponty, 1962, p. 143).[121] To get used to a stick, a feathered hat, a typewriter is to incorporate them into one's body. The incorporation of objects provides a means to expand the constricted lived spatiality of the patient. The physician can assist the patient to increase the range of bodily space by encouraging, where necessary, the habitual use of such things as visual aids, a cane, a wheelchair, a walker, and so forth. Often a patient may be reluctant to "give in" to such aids, or may view them as demeaning. If physicians and patients can learn to see such objects as extensions of bodily space (rather than as symbols of disability) they can utilize them effectively to increase the range of the patient's possible actions.

In this connection, it is interesting to note that physicians often pay little attention to exploring with patients such matters as the various means by which they can counteract the constriction of lived spatiality. For example in 1981 Dr. Dewitt Stetten, Jr. (1981, pp. 458–460) poignantly described his experience with physicians during a fifteen-year battle with macular degeneration. He noted that, as he struggled to cope with his progressive blindness, not one of the seven "distinguished and highly qualified" ophthalmologists he consulted at any time suggested any devices that might be of assistance to him; not one mentioned "any of the ways in which I could stem the deterioration in the quality of my life." To learn about such aids he had to depend upon friends and acquaintances who themselves had impaired vision.[122]

This difference in perspectives between physician and patient reflects, of course, their disparate systems of relevances. Since the physician interprets the patient's lived experience of illness in terms of the scientific understanding of anatomy, physiology, and so forth, he or she views the bodily dysfunction primarily in terms of likely medical interventions, whereas the patient regards the problem primarily in terms of the relief of lived body disruption. As Stetten's article emphasizes, however, it is vital for the physician to recognize that the distressing effects of illness may include such factors as a change in the character of lived spatiality. Most importantly, such factors can be directly addressed and relieved – even in the event that the possibility for successful medical intervention is limited.

It will be recalled that illness also engenders a change in lived temporality. The significance of past, present and future may take on a
different character such that the patient may be caught in the past
(obsessed with the meaning of past experiences), confined to the present
moment (preoccupied with the dictates and demands of the here and
now), or projected into the future (living in terms of what may happen). In
directly addressing this changed character of lived temporality, physicians
can do much to help patients address the problems associated with a
change in temporal significance. Past meanings and future fears can be
directly addressed in a realistic fashion, thus enabling the patient to live
more effectively in the present. In this connection, it is important to note
that fears for the future are almost always concrete. "Will I be able to
walk from my office to the classroom?" "Will I be able to sit at my desk
all day?" "Will I embarrass myself in a public gathering?" "Will my
illness be prolonged and prevent me from carrying out an important
project?" Once these fears have been explicitly recognized, strategies can
be evolved to deal with them. Paying attention to, and dealing with,
concrete fears is particularly important for patients with chronic, degenerative conditions. The loss of control which accompanies chronic illness can
often produce a sense of helplessness and a feeling of global uncertainty
about the future. Focusing on concrete fears enables the patient to regain
some control over his or her situation.

As noted earlier, gestural display is an essential feature of lived body.
Engel (1977a, p. 224) argues that gestural display is a major component
of communication which must be carefully attended to by the physician.
Indeed, gestures may indicate a different message from the spoken word.
As an example, he describes a hospitalized female patient who, on being
asked how she was doing by her physician, replied "Pretty good, I guess,"
but who, at the same time, frowned slightly and raised and then let fall her
right hand in the gesture of helplessness. The physician ignored the
gesture, replied "Good, I'm glad to hear that," and walked out of the
room, having indicated that he was discharging her from the hospital.
Noting that the patient appeared disconsolate, Engel remained behind and
commented that she did not seem too happy about her discharge,
whereupon she burst into tears and recounted to him some important
information about her personal life which had direct bearing on her
illness. She indicated that she had wanted to share this information with
her physician but that he had not given her the opportunity. The physician
in question later expressed surprise at the information and amazement at

how readily the patient had revealed it. Engel (1977a, p. 225) goes so far as to say that, not only is the inattention to gestural display poor clinical science (in that it ignores vital information relevant to the patient's illness and appropriate treatment), but he suggests that we can develop a scientific typology of gestures, postures and facial expressions and establish their relationship to inner experiences being felt and expressed.[123] In any event, it is clear that gestural display is an important component in communication which should be explicitly attended to.

It is also important to recognize that changes in gestural display may be a source of suffering for the patient. Indeed, feelings of shame are endemic among persons with disabilities. In this respect, it is interesting to recall Sartre's analysis of "the Look." Sartre argues that I experience my body as a "being-for-the-Other" in moments of shame and humiliation, even if no-one is actually looking at me. In such moments I see my body through the eyes of the Other. For persons with overt disabilities "the Look" is, more often than not, concretely experienced in the sense that others *are* viewing one's body in a negative fashion. Consequently, the sense of alienation from body which occurs in the experience of "being-for-the-Other" is particularly profound for those patients with a disordered body style.[124]

Leder (1984a, p. 36) has suggested that an understanding of embodiment as lived body (as being that which I AM rather than a passive, impersonal object) can motivate a sense of personal responsibility for bodily functioning. Such a sense of personal responsibility focuses attention on the role of personal participation in prevention of illness and treatment of disease. In this connection, Leder has noted that patients often simply hand over their bodies to the physician for treatment.[125] This is undoubtedly the case. In the clinical encounter the body becomes objectified. With this objectification the unity of lived body disintegrates and the body is alienated from the self. The alienation from self engenders a profound sense of loss of control. Illness deeply erodes one's sense of autonomy; one loses confidence in the ability effectively to manage one's physical situation. Hence one abrogates personal responsibility in favor of allowing the physician to assume control. This handing over to the physician, in and of itself, further adds to feelings of helplessness. Part of the healing function is to assist patients in reasserting their autonomy in the face of the disintegration of lived body. This implies paying explicit attention to the various disturbances in the patient's world and in the perceived change in the relation between self and body. In exploring with patients the particular impact that illness has upon their lives and discussing ways to minimize such impact, the physician can assist patients in asserting their selfhood. This is the case even in the event that such control is necessarily limited to the manipulation (rather than cure) of symptoms.

Moreover, the understanding of body as lived body effectively mitigates against some of the dehumanizing aspects of medical care. In conceiving of the body as an exclusively biophysiological mechanism, medicine in effect abstracts the body from the person whose body it is. The primary focus is on the physical disease process with a concurrent de-emphasis on the disorder of self and world. The patient may be seen as a "well controlled diabetic" or an "interesting carcinoma" rather than as a suffering subject. For the patient the disturbance of lived body (disruption of body/self/world) is of primary importance. If therapeutic intervention is concentrated solely on the dysfunction of the biological body, with little attention paid to the disturbance of lived body, patients feel themselves reduced to the status of a physical object, and consequently dehumanized. In addition, the denigration of subjective experiencing of disorder in favor of an exclusive preoccupation with "objective," quantitative clinical data, further adds to the patient's loss of personhood.

In this respect it is important to iterate the distinction made between suffering and clinical distress. In particular, it will be recalled that suffering is experienced by persons, not merely by bodies. Cassell (1990; 1991) argues that suffering occurs when the impending destruction of the *person* is perceived. Consequently, suffering relates not only to the loss of intactness of the biological body but to the loss of integrity of the whole web of interrelationships between body, self and world.[126] It seems clear, then, that suffering is intimately related to the disruption of lived body, to the manner in which one uniquely *exists* one's body, and to the disruption of that embodiment which alters all one's relations and interactions with the surrounding world. If suffering is to be alleviated, such disruption of embodiment must be directly addressed.

The analysis of the body-as-intended reveals that, in addition to the disruption of lived body experienced at the pre-reflective level, illness engenders a shift of attention which necessarily results in the objectification of one's body as a material object and (more particularly) as a malfunctioning physiological organism. Thus, the dialectic of identification and objectification is integral to the experience of illness. On the one hand at the immediate level of bodily disruption I AM my body and I

EXIST my illness; on the other, my impaired body demands my attention and thus I objectify it and experience a distance from it. The object-body is that body which I "have" or "possess" (rather than being simply the lived body which I "exist"); furthermore, it altogether escapes me being a purely biological body with its own nature. Consequently, I find myself alienated from it. This sense of alienation between body and self, which is intrinsic to the experience of illness, is intensified in the "medical gaze" with the reduction of body to the status of a scientific object.

The objectification of body results in the loss of embodiment. That is, as an object the body is no longer embodying. This loss of embodiment is part of the relevance structure of illness (incorporated into the finite province of meaning of those who experience illness). Consequently, sick persons can identify and recognize this loss of embodiment as an integral element of illness (i.e., they can share something of another's experience of illness without the need for physiological explanation). In particular, the sick have a mutual understanding of the manner in which the body is apprehended in illness (as an oppositional force, as a physical encumbrance, as a malfunctioning physiological organism, as that which is "uncanny," and so forth) which provides immediate recognition of another's circumstance.

It is important to note that this empathic understanding of the "givenness" of illness, founded on the apprehension of body-as-object, provides a clue as to the manner in which physician and patient can develop a shared world of meaning with regard to the patient's illness. Such an empathic understanding is available to all (even to those who have not experienced illness in their own lives). This is the case because, as we have seen, under normal circumstances the body is apprehended as an object in ways that point towards its apprehension as body-as-object in illness. From time to time, in everyday occurrences, one becomes aware of one's body as a material, physical entity, as a physical encumbrance, and as a physiological organism. Like the apprehension of body-as-object in illness, this mundane experience of the object-body is one of alienation (of a separation between body and self). Furthermore, in this common occurrence the body may appear as "uncanny." The lifeworlds of physician and patient thus provide the starting point for mutual understanding (Husserl, 1970b, p. 255; Engel, 1985, p. 364). The "givenness" of illness is available to the physician not merely through a subjective experience of being sick (although obviously this provides greater insight), but through reflection upon the manner in which the body is experienced as an object in everyday life.[127]

The analysis of the body-as-object also suggests that, whereas the objectification of body necessarily results in the alienation of body from self, different bodily dysfunctions have differing impacts on the relation between body, self and world. For example, as has been noted, the experience of the body as a physical encumbrance is most evident in those illnesses involving an overt loss of function. Indeed, in another context, I have argued that illnesses may be understood in terms of the differing existential meanings which relate to the varied bodily disturbances (Toombs, 1991). For example, motor disturbances produce a bodily alienation through the loss of corporeal identity and the establishment of the body as an *oppositional force* which is beyond the control of the self. The disruption of motor function diminishes one's capacity to act within the world in an essential way. In consequence, the world itself assumes an unusually problematic and restrictive character. On the other hand, disturbances involving the loss of sensation precipitate a radical disengagement of body from self in that the body is no longer experienced as "mine" or as "belonging to me." With disruptions such as bowel and bladder disorders (which represent the most elemental loss of control over one's body), the body is experienced not merely as oppositional but as malevolent, posing a constant threat to one's dignity and self esteem. The foregoing analysis of the body suggests, therefore, that it is possible to identify the differing impact of various bodily disorders upon the patient's manner of being-in-the-world.

In addition, the process of identification with or disassociation from one's body (which is an important element in illness) varies according to the type of bodily disorder. In chronic illnesses patients cannot disassociate themselves from their diseased bodies and consequently they find themselves inescapably embodied, irrevocably attached to an essentially malfunctioning bodily organism which promises to disrupt all their involvements in the world. Such disorders are consequently, experienced as profoundly *world threatening* (even if not life threatening). In acute disorders, such as appendicitis, disassociation from the diseased body part is not only possible but, in many cases, advisable. An awareness of the existential meanings associated with particular bodily disruptions and differing disease processes can provide invaluable insight into the patient's lived experience of illness.

# CHAPTER FOUR

# THE HEALING RELATIONSHIP

The foregoing phenomenological analysis has demonstrated that illness and body mean something significantly and qualitatively different to the patient and to the physician. This difference in perspectives is not simply a matter of different levels of knowledge but, rather, it is a reflection of the fundamental and decisive distinction between the lived experience of illness and the naturalistic account of such experience. It has been further noted that this difference in understanding between physician and patient is an important factor in medical practice – a factor which has an impact not only on the extent to which doctors and patients can successfully communicate with one another on the basis of a shared understanding of the patient's illness but, additionally, on the extent to which the physician can address the patient's suffering in an optimal fashion.

If the physician is to minimize the impact of this difference in perspectives and develop a shared understanding of the meaning of illness, it is vital that he or she gain some insight into the patient's lived experience. It should be emphasized that attending to the patient's lived experience is important for a number of reasons: (1) the physician's appreciation of the patient's lived experience of illness is important for the physician acknowledging the patient as a person and treating the patient as a person; (2) understanding the lived experience is necessary in order that the physician may then interpret such understanding in terms of his or her knowledge of anatomy, physiology, and so forth, and begin the process of therapeutic intervention; (3) an adequate understanding of the lived experience is important for insuring the most effective scientifically mediated therapeutic intervention; and (4) the act of healing requires that physician and patient share a common understanding of the patient's illness.

In this chapter I shall explore some ways in which it is indeed possible for the physician to gain insight into the lived experience of illness and, thereby, to establish a shared world of meaning with patients. In particular, I shall suggest that (1) illness-as-lived may be understood as a particular way of being which exhibits certain typical characteristics; (2) the manner in which the body is apprehended under normal circumstances provides a clue for developing an empathic understanding of illness; and (3) the clinical narrative is an important element in disclosing what illness means to a particular patient. Finally, I shall discuss the nature of the physician-patient relationship (the healing relationship) and argue that the act of healing *requires* an understanding of illness-as-lived.


## 1. ILLNESS-AS-LIVED[128]

The phenomenological description of the manner in which illness is apprehended in the "natural attitude" has disclosed that illness means much more to the patient than simply a collection of physical signs and symptoms which define a particular disease state. Illness is fundamentally experienced as a global sense of disorder – a disorder which includes the disruption of the lived body (with the concurrent disturbance of self and world) and the changed relation between body and self (manifested through objectification and alienation from one's body).

Further reflection upon this global sense of disorder which *IS* the lived experience of illness in its qualitative immediacy, discloses that the lived experience exhibits a typical way of being – a way of being which incorporates such characteristics as a loss of wholeness, a loss of certainty, a loss of control, a loss of freedom to act, and a loss of the familiar world. Such characteristics are intrinsic elements of the illness experience, regardless of its manifestation in terms of a particular disease state.

Illness is primarily experienced as a fundamental *loss of wholeness*, a loss of wholeness that manifests itself in several forms. Fundamentally, of course, this loss of wholeness manifests itself in the awareness of bodily disruption or impairment – an awareness that is not so much a simple recognition of specific symptoms (e.g., shortness of breath) as it is a profound sense of the loss of total bodily integrity. The body can no longer be taken for granted or ignored. It has seemingly assumed an opposing "will" of its own, beyond the control of the self. Rather than functioning effectively at the bidding of the self, the body-in-pain or the body-malfunctioning thwarts plans, impedes choices, renders actions impossible. Illness disrupts the fundamental unity between the body and self. As Cassell notes:

Disease can so alter the relation (with one's body) that the body is no longer seen as a friend but, rather, as an untrustworthy enemy. This is intensified if the illness comes on without warning, and as illness persists, the person may feel increasingly vulnerable (Cassell, 1982, p. 640).

In illness the body, which was hitherto simply lived, becomes the unwelcome object of one's attention. This objectification necessarily results in a sense of alienation from one's body. In particular, the body is experienced as no longer embodying. Rather, it manifests itself as if a material physical object or as an oppositional force which must be overcome in carrying out one's projects.

In addition, the malfunctioning object-body reveals itself as a hidden and alien presence which is essentially beyond one's control. The perceived disruption in function discloses the machine-like nature of the biological body and the various and varied physical processes which are neither directly experienceable nor controllable. This sense of "otherness" of body is acutely felt by the patient in discussions with the physician. The biological, pathological sense of the body is of the body as other-than-me, of the body in opposition to the self, and it is this sense that is now emphasized.

Even if the body is eventually restored to health, the perceived loss of bodily integrity remains, especially if the illness is serious.

Having a heart attack is falling over the edge of a chasm and then being pulled back. Why I was pulled back made no more sense than why I fell in the first place. Afterward I felt always at risk of one false step, or heartbeat, plunging me over the side again. I will never lose that immanence of nothingness, the certainty of mortality. Once the body has known death, it never lives the same again.

People who think of themselves as healthy walk that edge too, but they see only the solid ground away from the chasm (Frank, 1990, p. 16).

Illness forces one to recognize in an explicit way the tenuous nature of bodily integrity and the lack of control one has over bodily functioning.[129] Obviously, this is most evident in chronic and life-threatening illness. But it is also an intrinsic element of less serious bodily disorders. Having once experienced illness, one recognizes that one can no longer take the body's future compliance completely for granted.

The loss of bodily integrity incorporates not only an awareness of alienation from, but also unwilling identification with, one's body. While on the one hand there is an apprehension that the body is other-than-me and thus essentially out of my control, on the other hand there is an acute awareness of inescapable embodiment. That is, one understands oneself to be inextricably embodied in that one cannot totally disassociate the self from this malfunctioning biological body which promises to disrupt all one's involvements in the world.

In this regard it is very important to note that the loss of wholeness experienced in illness manifests itself not only as a threat to the body but also as a concrete threat to the self.

Sickness ... shatters the web of assumptions on which our lives are based. We take it for granted that our arms, legs, fingers, feet and other organs will respond to our commands. When they do not – when we cannot move as we wish – we discover how much of our sense of self is bound up with our body and how disoriented we become when that body turns into our enemy rather than ally (Silberman, 1991, p. 13).

In this respect I would refer the reader back to the phenomenological description of the lived body as one's being-in-the-world. Since the lived body, as being-in-the-world, is the locus of one's intentions, it is not surprising that the disruption of lived body occasioned by illness is experienced as a disruption of self.

The disintegration of self is particularly acute in the experience of serious illness since one's relationships with others, one's work, one's sense of who one might become, one's sense of what life is and ought to be, all change and the change is "terrifying" (Frank, 1991, p. 6).

Robert Murphy reflects this loss of self when he writes:

From the time my tumor was first diagnosed through my entry into wheelchair life, I had an increasing apprehension that I had lost much more than the full use of my legs. I had also lost a part of my self. It was not just that people acted differently toward me, which they did, but rather that I felt differently toward myself. I had changed in my own mind, in my self-image, and in the basic condition of my existence.... (I)t was a change for the worse, a diminution of everything I used to be.
*Disability is not simply a physical affair ... it is our ontology, a condition of our being in the world (Murphy, 1987, pp. 85, 90).* (Emphasis mine).

It is important for physicians to recognize the primacy of the threat to the self which is occasioned by illness, especially since "fixing" the body does not necessarily alleviate the threat to the self.

The loss of wholeness not only incorporates a perception of bodily impairment and loss of integrity but also includes the *loss of certainty* in its most profound form. In the experience of illness one is forced to surrender one's most cherished assumption – the assumption of personal indestructibility. And if this most deeply held assumption is no more than an illusion, what else in one's hitherto taken-for-granted existence can remain inviolable?

We take it for granted ... that life is predictable and we are immortal and that we can, therefore, control our own fate.... (I)llness destroys this primordial sense of invulnerability, forcing us to acknowledge our impotence and our mortality (Silberman, 1991, p. 13).

Once shattered, the illusion of personal indestructibility can be only tenuously re-established.

The radical loss of certainty that accompanies illness is cause for considerable personal anxiety and fear. Although often acutely conscious of being afraid, persons who are sick find it difficult to communicate this apprehension to others. Paradoxically, patients often deem such concern to be inappropriate even though it is ineluctably part of the experience of illness. In attempting to minimize the patient's anxiety, the physician may make an effort to discuss the illness or therapeutic intervention in such a way as to imply that there is no real cause for concern. The patient, however, may interpret this simply to mean that the experience of anxiety is therefore irrational and inadmissible.[130]

For the most part illness is experienced as a capricious interruption, an unexpected happening, in an otherwise more or less carefully formulated life plan. The disease is perceived as "befalling the person, as an unasked-for and unanticipated 'happening-to-me,' falling outside the person's range of possible choice and plans" (Zaner, 1982, p. 50). And thus, accompanying the profound sense of loss of wholeness and loss of certainty, is an acute awareness of *loss of control*. The familiar world, including the self, is suddenly perceived as inherently unpredictable and uncontrollable. Illness, as Pellegrino (1982, p. 159) has noted, "moves us ... toward the absorption of man by circumstance."

It is, of course, the case that in the effort to give explanatory meaning to an illness, the patient may alternatively associate sickness with some sort of punishment (divine or otherwise). That is, the person who develops a serious illness may feel that he or she *must* have done something to deserve it and that the illness is, thus, a retribution for wrongdoing.[131] Nevertheless, even in the event that illness is perceived as the result of personal transgression or inappropriate behavior, rather than simply as an inexplicable random interruption of one's life plan, the individual is still acutely aware of a fundamental loss of control over the *present* situation. He or she cannot *now* undo the transgression which is perceived to have caused the distress, cannot *now* change the present circumstance.[132]

The loss of control which is intrinsic to illness is acutely felt by modern man in light of the illusions we harbor about the power of technology and the capabilities of modern science. Since technology and science have been extremely successful in eradicating or ameliorating many diseases, not only is illness perceived as an unwarranted intrusion but the sick person expects medical intervention to provide nothing less than a complete restoration of health. The patient thus comes to the physician with the unrealistic expectation that such a complete restoration of health will be forthcoming. If the physician is unable to fulfill this expectation, the patient is overwhelmed by a sense of helplessness and perceives the situation to be totally and irrevocably out of control.[133]

The technology that promises redemption paradoxically intensifies the loss of control experienced in illness. When undergoing investigation patients find themselves at the mercy of faceless machines – machines with barely understood functions but whose dictates must be obeyed. In this encounter with machines one perceives oneself to be an object of investigation, rather than a suffering subject. This transformation to objecthood is concretely felt not only in the "gaze" of machines, but also in the "gaze" of health care professionals.[134] In the transformation to objecthood, the patient feels no longer able effectively to control what happens.

The loss of control also manifests itself concretely in the experience of having to rely on others to do what one has formerly been able to do for oneself. Illness, in its various forms, always impedes the ability to be self-reliant, to act on one's own behalf.[135] The ill person must not only seek the help of others for physical assistance but must also rely upon the help of a trained healer, a physician. This relationship is an inherently unequal relationship in that the physician "professes to possess precisely what the patient lacks: the knowledge and power to heal" (Pellegrino, 1982, p. 159). There is an asymmetry of power in favor of the physician (or other health care professional) in view of his or her knowledge of disease, technical skills, position of authority, and so forth. The inequality of the relationship accentuates the loss of control felt by the ill person.

Illness also erodes the capacity to make rational choices regarding one's personal situation because the one who is ill:

[D]oes not understand what is wrong, how it can be cured, if at all, what the future holds, or whether the one who professes to heal can in fact do so. The ill person has not the knowledge or skills requisite for curing his own bodily or mental illness or to gain relief from his pain or anxiety. His freedom to act as a person is severely compromised (Pellegrino, 1982, p. 159).

When competent, patients are ultimately responsible for making clinical decisions with regard to their care. Nevertheless, although such decisions are usually made after appropriate advice and consultation with the physician, the sick person almost always feels inadequate to the task. The decision is uniquely personal, not only in that it involves a personal responsibility for choice but in that this decision will ultimately affect one's unique life plan. Although aware of the necessity for choice, patients often feel that they do not possess the knowledge or the capacity to make the decision in a rational manner. Sometimes they may intuitively feel that the course of action recommended by their physician is not in their best interests and yet – more often than not – patients do not feel free to reject the advice of the physician. To do so would seem to be irrational in the face of the inadequate knowledge they feel themselves to possess. To do so would also be to risk alienating themselves from the one who promises to alleviate their distress.

This difficulty remains even when the patient is a physician since knowledge about one's illness and the eventual outcome of clinical decisions is necessarily incomplete.[136] There are large gaps in understanding for even the best-understood diseases. Causes may be obscure and the outcome is always a matter of probability rather than the certainty which the patient seeks.[137]

Furthermore, illness erodes the capacity to make rational choices in another respect. Illness impairs the ability to reason in that it is difficult to be clear-headed when one is suffering or in pain (Cassell, 1977, p. 17). It is also extremely hard to be "clear-headed" and view clinical choices with equanimity when they relate directly to one's *own* uncertain future.

In reflecting upon what is in his or her own best interest, the individual does so in light of a life plan and a unique system of values. Each person lives life according to certain fundamental principles that have personal meaning, and it is in light of these principles that one makes choices and acts in the world of everyday life. In the existential crisis of illness, these fundamental personal values are often made explicit. The individual encounters and interprets the threat to the self by reference to, and in light of, the principles which render his or her life meaningful.

Invariably patients assume (often incorrectly and certainly un-reasonably) that their physician knows and understands what this personal value system is and, further, that in making the clinical decision the physician is doing so not only in light of the clinical data but additionally with regard to this personal value system. Patients, therefore, rarely explicitly communicate their values to their physicians. Physicians, on the other hand, may deem it inappropriate, irrelevant, or intrusive to ask patients about their values and may judge the clinical data alone to be sufficient to determine what is in the patient's best interest. Thus, patients not only lose the freedom to make a rational choice regarding their personal situation but additionally lose or abrogate the freedom to make the choice in light of a uniquely personal system of values.[138]

Illness is a state of disharmony, disequilibrium, dis-ability, and dis-ease which incorporates a *loss of the familiar world*. As the phenomenological analysis has revealed, illness represents an altered state of existence, a distinct change in one's being-in-the-world. This change may be tem-porary (as with a bout of influenza) or long-lasting (as with a chronic disorder). In this altered state the sick person is unable routinely to carry on normal activities, to participate in the everyday world of work and play. This isolation is all the more acute because the familiar world continues on as normal. Associates pursue their activities much as they have in the past, and although illness affects the totality of a sick person's own experiencing, it is a "fact" that is necessarily only in the periphery of the experience of others.

This point is powerfully illustrated in the following passage from *The Death of Ivan Ilych*. Ilych arrived home from the doctor's office and:

[B]egan to tell his wife about it. She listened, but in the middle of his account his daughter came in with her hat on, ready to go out with her mother. She sat down reluctantly to listen to his tedious story, but could not stand it long, and her mother too did not hear him to the end.... There was no deceiving himself: something terrible, new, and more important than anything before in his life, was taking place within him of which he alone was aware. Those about him did not understand or would not understand it, but thought everything in the world was going on as usual. That tormented Ivan Ilych more than anything (Tolstoy, 1978, pp. 521, 524).

Existential aloneness is necessarily a part of serious illness. As Murphy (1987, p. 63) remarks, "Nothing is quite so isolating as the knowledge that when one hurts, nobody else feels the pain; that when one sickens, the malaise is a private affair; and that when one dies, the world continues with barely a ripple."

The taken-for-grantedness of everyday life is disrupted, not only in the sense that routine activities and involvements are disturbed (and become "problematic"), but additionally in the sense that the usual experience of time and space undergoes a significant change. The unavoidable preoccupation with pain, sickness, or incapacity, grounds one in the present moment. Illness truncates experiencing. The future (long or short-term) is suddenly disabled, rendered impotent and inaccessible.[139]

Moreover, the disruption of one's embodied capacities necessarily results in the constriction/restriction of surrounding space. Space constricts to the extent that illness confines (to the bed, to the house, to the hospital). Furthermore, with loss of function, physical space itself assumes a restrictive character (a location may be "too far" from the bed or chair, a step "too high" to climb, a room "too crowded" to navigate). Thus, the hitherto familiar world is permeated with a global sense of disorder. It is a world in which one is no longer at home.

This "unfamiliarity" also manifests itself in a changed response to one's environment. In sickness the world "impinges upon" the senses in a qualitatively different manner. Sights, sounds, smells, tastes, touches, associations with others – all may take on a different significance in illness. Indeed, the very way in which the "world" as a whole appears (joyful, drab, inviting, repelling, intrusive, and so forth) is affected by the transformation from health to sickness.

## Summary

In sum, then, the experience of illness represents a distinct way of being in the world, a way of being that is typically characterized by a loss of wholeness and bodily integrity, a loss of certainty and concurrent apprehension or fear, a loss of control, a loss of freedom to act in a variety of ways, and a loss of the hitherto familiar world.[140]

As a first step towards achieving a shared understanding of the meaning of illness, physicians can learn to recognize, and pay attention to, these typical characteristics of the human experience of illness. This means temporarily setting aside the naturalistic interpretation of illness in terms of theoretical disease constructs, in order to focus explicitly upon illness as it is lived through by the patient. This is not intended to imply

that the physician is thereby required to "give up" his or her understanding of illness as a disease state. Rather, it is to suggest that he or she
perform a temporary "shift in consciousness" from a purely "naturalistic"
construction of the patient's disease to a lifeworld interpretation of the
patient's disorder in order to gain a more complete understanding of the
patient's illness. This shift in focus will not only provide insight into the
human experience of illness but will enable the physician to address the
patient's suffering more directly.

## 2. EMPATHIC UNDERSTANDING

As was noted in the previous chapter, those who have been sick share an
empathic understanding of the "givenness" of illness in that they have a
mutual understanding of the manner in which the body is apprehended in
illness – as an oppositional force, a physical encumbrance, as that which
is "uncanny," and so forth. Consequently, sick persons can share something of another's experience of illness regardless of the disease state and
without the need for any physiological explanation. It was further
suggested that this empathic understanding of the "givenness" of illness,
founded on the lived experience of the body, provides a clue as to the
manner in which a shared world of meaning may be established between
physician and patient.[141] In this section I shall explore further the notion
that the lifeworlds of physician and patient do indeed provide the starting
point for mutual understanding with regard to the illness experience and
suggest that reflection upon the manner in which the body is apprehended
in normal circumstances can provide important insights into illness-as-
lived.

The phenomenological analysis has revealed that the experience of
illness renders explicit the "ambiguity" of being at once identified with
one's body (in that one is embodied) and yet separated from it (in that it is
essentially out of one's control). This awareness of the "ambiguity" of
one's body is not, however, limited to those moments when we become
sick *although, of course, such moments have undoubtedly a greater
existential import with regard to the significance we attribute to such
ambiguity.* In the course of everyday, mundane existence the recognition
of such ambiguity periodically breaks in to consciousness. We are
reminded of the fact that we are embodied (and that we cannot disassociate ourselves from our bodies) whenever, for example, physical

limitation prevents us from accomplishing a task or pursuing an ambition, in the inevitable experience of growing older, in the approving (or disapproving) glance of the stranger as we walk by, in the habitual awareness that this is "my" body and that while I claim it as my own, I am conjoined with it in a *symbiotic* relationship such that I cannot dispense with it altogether if *"I"* am to remain. On the other hand, mundane experiences also remind us that the body is essentially Other-than-me in that it is a material, physical object over which I have, at best, very limited control.

For the most part such experiences in everyday life are typically easily passed over and forgotten. However, they provide valuable clues as to the sense of disorder which is intrinsic to illness. In reflecting upon those moments when I recognize my inescapable embodiment – for example, experiences of physical limitation or the regretful apprehension I may have from time to time that my body attests to the fact that I am "no longer as young as I used to be" – I may grasp something of the "shock" of radical bodily limitation which serious illness forces upon the sick (a "shock" which persists while illness lasts and which incorporates the awareness of loss of control).

In particular, everyday experiences of the body as a material physical object can disclose the bodily alienation manifested in illness. As has been noted, in such ordinary occurrences as my leg "going to sleep," the body is experienced as no longer embodying but rather as "thinglike" and separate from the self. Such an everyday occurrence is hardly distressing at a deep level. All that is required is that I massage the leg in order to "enliven" it and "gain it back." But a little reflection on this experience suggests how much more distressing it would be if such "deadness" were to persist indefinitely.

Furthermore, in one way or another, we all have mundane experiences of the body as an oppositional force – a force which must be overcome in order to carry out our projects in the world. If I am fatigued for lack of sleep or due to "jet lag," I must fight my body's reluctance to participate in my efforts to be wakeful, invigorated and enthusiastic ("enlivened"). If I have imbibed too freely (and have a "hangover"), or feasted too unwisely (and have a stomach-ache), I must surmount my body's resistance if I am to eat the breakfast which my host or hostess has carefully prepared and put before me. Such experiences place the body in opposition to the self, reveal the body as a physical encumbrance. Again, such mundane occurrences are, for the most part, only temporarily disquieting. Nevertheless, one can readily appreciate how much more disquieting it would be if such occurrences were to persist; if, as in illness, the body as an oppositional force was an ever-present reality.

Additionally, experiences in which one's body is recognized as a mechanical physical object are not limiting cases confronted only in unusual circumstances. As Engelhardt (1973, p. 38) has pointed out, one has everyday confrontations with the automaticity of one's reflexes – one bangs one's patella and the knee jerks; one drinks too much coffee and the heartbeat becomes irregular. In such experiences the body presents itself as an other, as an object among others to be felt, seen and acted upon. Moreover, the body presents itself as a biological body with its own nature, functions, and physiological processes of which one has little, if any, awareness and over which one has at best very limited control. Under normal circumstances such experiences of the body as a physiological entity may not be particularly threatening. One may, for example, simply wonder at the marvelous complexity of the workings of one's mechanistic body. Nevertheless, such everyday experiences do necessarily reveal the body's "uncanniness" – in particular its hidden and alien presence – and, consequently, they do remind one of the tenuous nature of the control which one exercises over the physical body (a reminder which is oddly disquieting).

Some reflection on this apprehension of the body as "uncanny" under normal circumstances provides a clue as to the profound sense of bodily alienation which is intrinsic to the experience of illness. It will be recalled that with bodily disorder not only is the body apprehended as a mechanistic physical entity but, more particularly, as a *malfunctioning* physical entity. As a malfunctioning physical entity the body is not only disclosed as hidden and alien presence in an overt and persistent manner but, additionally, in a manner which is necessarily perceived as threatening to the self. How much more disquieting it must be in these circumstances to apprehend the body's independent nature and to recognize the extent to which one has but little, if any, control over it.

It seems clear, then, that the lived experience of the body is the starting point for mutual understanding with regard to the illness experience. Whether or not the physician has personally experienced sickness, from time to time under normal circumstances he or she apprehends the body's ambiguity, senses its hidden and alien presence. Some reflection on such mundane occurrences in everyday life provides the basis for empathic understanding of the "givenness" of illness. Such understanding is not esoteric. We all have everyday experiences of the body as an oppositional force, as a physical encumbrance, as a material, physical object – experiences which alienate self from body and which disclose the body as "uncanny." What distinguishes these experiences from their counterpart in illness is that in everyday life they are short-lived and, for the most part, easily forgotten. Nevertheless, they point towards the deep sense of bodily alienation and loss of control which is intrinsic to illness-as-lived.

Not only do the lifeworlds of physician and patient provide the basis for empathic understanding of the experience of being ill but the lifeworld is also the basis for comprehending the lived experience of specific bodily disorders. Indeed, as Engelhardt (1973, p. 38) has noted, medical students learn explicitly to observe their own life processes in order to notice the dimension of the purely physical and mechanical.[142] The physician achieves a scientific understanding of illness on the basis of his or her pre-scientific experience of everyday life. Engel (1985, pp. 362–366) argues that in attending to the patient's report of disorder, the physician's response is predicated on a personal lifeworld and lived body. To become breathless, for instance, is part of a common lifeworld for both patient and physician, as is the panic engendered when breathlessness occurs for no apparent reason.

Cassell (1985a, pp. 46–47) suggests that it is vital for the physician to use personal experience of body and world as a framework of reference to help understand the complaints of patients. In doing so, he argues, the physician is checking with his or her own knowledge to be sure that a symptom is completely grasped. The physician uses personal experience of the lived body to check what a patient means and to grasp how the symptom differs from normal function, and, in addition, uses the knowledge of anatomy and physiology to guide the process of inquiry.[143] In the event that one has never had a similar experience or felt the body sensation described by the patient (a problem more often encountered perhaps by the young and inexperienced) then, says Cassell (1985a, pp. 46–47), one has to ask more questions so that "when you have finished you have both acquired the diagnostic information and learned more about the world."

Engel notes that by virtue of his or her education the physician's lifeworld is expanded.

The report by a dyspneic patient, for example, resonates not just with the physician's life-long awareness of breathing and breathlessness – his/her own and others' – but also with his/her more recently acquired experience with lungs and blood gases and unsaturated hemoglobin and audible rales and patients struggling for air. To the extent that these products of formal education constitute lived experiences rather than abstractions learned from books, they contribute to the ever-changing lifeworld of the physician (Engel, 1985, p. 364).

Engel makes the important point that the lifeworld of the patient also evolves in the course of illness and health care and in the process there comes about a mutual understanding between physician and patient in the context of which ongoing care takes place.

## *Summary*

In sum, then, physician and patient share lifeworld experiences which provide the basis for a shared world of meaning between them. In particular, some reflection upon the lived experience of the body suggests that an empathic understanding of illness-as-lived is readily available to the physician – even if he or she has not personally experienced sickness.[144] This empathic understanding of the "givenness" of illness founded on the lived experience of the body under normal circumstances may be enriched to the extent that the physician regularly observes those who are living with illness. Obviously, such empathic understanding may also be further enriched by a personal experience of sickness. As Sacks notes with regard to his own leg injury:

[G]oing through all the specific experiences of "The Leg," as well as the more general experiences of "Being a Patient," taught me, changed me as nothing else could. Now I *knew*, for I had experienced myself. And now I could truly begin to understand my patients.... I could listen to them, I could understand them, and sometimes I could help, because I had traversed this region myself (Sacks, 1984, pp. 202–203).

The phenomenological analysis of body reveals, however, that whereas such a personal experience indeed provides profound insights into the disordered existence of illness, more mundane lived experiences also provide invaluable clues as to the meaning of illness-as-lived.

### 3. CLINICAL NARRATIVE

The foregoing analysis has revealed that, not only is the lifeworld interpretation of illness distinct from the naturalistic interpretation of the disease state, but the meaning of illness is intimately related to the patient's unique biographical situation. In this section I shall argue that clinical narrative (the story of the illness as told by the patient) provides insights into the lived experience of illness and particularly into the meaning that illness has for a particular patient.

The clinical narrative is to be distinguished from the medical history. The medical history is based on the biomedical view of reality, the naturalistic interpretation of the disease state. Larry and Sandra Churchill (1989, p. 1127) have noted that the medical history concerns facts such as "onset of symptoms, disease etiology, pathophysiology, course of the disease, potential for and options for treatment." It also includes such items as inoculations, past injuries and hospitalizations, known allergies, and chronic conditions. The clinical narrative, on the other hand, is the story of the illness from the patient's point of view. It not only contains the facts of the illness (the story of the events which have brought the patient to the doctor) but, in addition, the patient's explanations, interpretations and understanding of such facts (Cassell, 1985a, p. 22). In particular, the clinical narrative incorporates a description of the manner in which the bodily disruption manifests itself in the life of the patient. The narrative (or tale) is "clinical" insofar as it lends itself to clinical or medical analysis. It is a narrative insofar as it is the patient's story of how he or she has been feeling, what he or she has been experiencing, in the realm of illness (Sacks, 1986, p. 14). In this respect Cassell (1985a, p. 15) points out that illness stories are different from other stories in that they almost always include two characters – the person and the person's body – to whom things happen.

In analyzing medical interviews between physician and patient, Elliott Mishler (1984, p. 14) has noted that there are two frameworks of meaning characterizing the discourse: the "voice of medicine" (representing the technical-scientific assumptions of medicine) and the "voice of the lifeworld" (representing the natural attitude of everyday life). The medical history reflects the "voice of medicine," the clinical narrative represents the "voice of the lifeworld." Mishler reports that in standard interviews the "voice of medicine" predominates. The physician controls the form and content of the interview, defining what is and is not

considered relevant through the questions he or she asks. While the
"voice of the lifeworld" periodically intrudes into the interview, the
physician quickly reintroduces the "voice of medicine" focusing on
"objective" symptoms in consonance with the biomedical model of
disease. Indeed, Mishler (1984, pp. 70–90) observes that physicians tend
to treat the "voice of the lifeworld" as non-medically relevant and
therefore quickly suppress this voice in the typical interview.[145] The
physician listens to the patient's narrative and extracts or abstracts from it
a (syndromic or etiological) "case" – based on his or her knowledge of
other "cases" and of the physiological and pathological processes of the
body (Sacks, 1986, p. 14).

It is important to note, however, that the "voice of the lifeworld"
directly relates to the patient's lived experience of illness. Consequently,
the clinical narrative – as opposed to the medical history – discloses what
illness means to the patient. As Cassell (1985a, p. 22) points out, patients
are not objective observers "reporting on the march of disease through
their bodies," rather they are instead telling about things that have
happened to them – happenings to which they assign meanings, interpretations and causal explanations which are a direct reflection of a particular
lifeworld. Thus, in the clinical narrative, the patient emphasizes what is
personally significant about the illness and its life impact.[146]

Attending to the patient's story is vital if one is to understand the
patient's illness. The assignment of meaning and explanation is as much a
part of the illness as is its physical expression. Just as illness without a
cough is different from an illness with this symptom, so an illness in
which the patient suspects or is afraid of, say, kidney involvement is
different from an illness without such fear (Cassell, 1985a, p. 126). Only
the patient's narrative can disclose such meanings.[147]

To gauge fully the patient's perspective and to be sure about what the
patient wants from care), the practitioner must elicit the patient's
explanatory model (as opposed to the biomedical explanatory model)
(Kleinman, 1988, p. 239).[148] This means explicitly attending to the
patient's narrative and also asking the patient to elaborate on such matters
as the reasons for the onset of symptoms at a particular time, his or her
lay understanding of what gave rise to the symptoms, and the expected
course and perceived seriousness of the illness. In addition, in order to
understand the meaning the illness has for the patient, Kleinman (1988,
p. 239) suggests the practitioner ask, "What is the chief way this illness
(or treatment) has affected your life?" and "What do you fear most about

this illness (or treatment)?" It should be noted with regard to the latter questions, that such matters are often disclosed by the patient in the course of the story of the illness. In particular, in providing a narrative account of the illness experience, the patient focuses on the lifeworld disruption which illness represents and tries to convey the impact of such disruption.

In providing a lifeworld description of bodily disruption, the clinical narrative deepens the physician's understanding of the lived experience of illness. Such narratives disclose what it is *like* to suffer from, say, multiple sclerosis, heart disease, arthritis. Clinical narratives do not operate at the level of objective pathology. They are not primarily concerned with such matters as elevated blood counts, demonstrable lesions, or abnormal EKG's. They detail the patient's experience of lived body disruption – of the disorder of body, self and world that illness represents.

As Sacks (1984, p. 202) has noted with regard to empathic understanding, it is difficult to "imagine" the lived experience of diseases such as Parkinsonism, because such experience is so far removed from normal everyday functioning. Nevertheless, a patient's description of what it is like to have Parkinson's disease can broaden one's understanding immeasurably. In his book, *Awakenings*, Sacks (1983) reports how he asked his patients with Parkinsonism to explain what life was like for them. In their stories they tell of profound disruptions of space and time – descriptions that can be found in no textbook definition of the disease state.

For instance, "Miss R" describes her experience as follows:

"What are you thinking about Rosie?"
   "Nothing, just nothing."
   But how can you possibly be thinking of nothing?"
   It's dead easy, once you know how."
   *"How exactly do you think about nothing?"*
   "One way is to think about the same thing again and again. Like 2 = 2 = 2 = 2; or I am what I am what I am.... It's the same thing with my posture. My posture continually leads to itself. Whatever I do or whatever I think leads deeper and deeper into itself.... And then there are maps. I think of a map; then a map of that map; then a map of that map of that map, and each map perfect, though smaller and smaller.... Worlds within worlds within worlds within worlds. Once I get going I can't possibly stop. It's like being caught between mirrors, or echoes, or something. Or being caught on a merry-go-round which won't come to a stop" (Sacks, 1983, p. 69).

As Howard Brody (1987, p. 96) comments with regard to such descriptions, "The richness of the patients' responses suggests how much we may still be missing with regard to many other diseases, where no careful listener has yet come along to hear the stories of the sufferers."

This underscores the importance of asking the patient to recount his or her story in detail, of posing such questions as "What is it like for you?" Surprisingly, perhaps, such questions are often not asked by physicians. As Murphy notes, for example:

Nobody has ever asked me what it is like to be a paraplegic – and now a quadriplegic – for this would violate all the rules of middle-class etiquette.... Polite manners may protect us from most such intrusions, but it is remarkable that physicians seldom ask either. They like "hard facts" obtainable through modern technology or old fashioned jabbing with a pin and asking whether you feel it. The tests supposedly provide good "objective" measures of neurological damage, but ... they reduce experience to neat distinctions of black and white and ignore the broad range of ideation and emotion that always accompanies disability (Murphy, 1987, p. 87).

Since medical people have a "penchant for looking primarily at the biological aspects of health," says Murphy, their advice often "dooms the patient to social and psychological disability in the name of somatic 'health,' whatever that is." Accordingly, some medical people consider a paralytic to be doing well "if he has no skin breakdowns, is not visibly depressed, and has clear bowels and bladder" (Murphy, 1987, p. 185).

Similarly, I have noticed that no physician has ever inquired of me what it is *like* to live with multiple sclerosis or to experience any of the disabilities that have accrued over the past seventeen years. Perhaps, most surprisingly, no neurologist has ever asked me if I am afraid or, for that matter, even whether I am concerned about the future. Yet, concern for the future is necessarily an integral part of the experience of chronic progressive illness. Indeed, such concern may be one of its most debilitating features.

If one is to understand the lived experience of illness, comprehend what the disorder means to the patient (and thus address directly the patient's disorder and suffering), then it is clear that one has to go beyond objective, quantifiable, clinical data and elicit the patient's illness story.[149]

Obviously clinical narratives are not limited to the stories individual patients tell in face-to-face encounters with their own practitioners. Insights about the lived experience of illness may also be gained from published narratives. That is, one may develop a greater understanding of illness-as-lived through literary accounts of the illness experience. All of us who read Sacks' (1983) report of his patients' descriptions of what it is like to have Parkinson's disease will gain some insight into Parkinsonism. Murphy's (1987) firsthand account of his ongoing experience of a spinal cord tumor and the resulting progressive paralysis provides his readers with concrete information about the disruption of self and world which such a disorder represents.

Baron (1985, pp. 606–611) argues that, indeed, literary accounts of illness may, in a sense, be read as "medical treatises that give physicians information absolutely essential to the practice of medicine." Such accounts contrast markedly with the biomedical description of disease.[150] Consider, for example, the following account of a patient with skin disease:

Oct. 31. I have long been a potter, a bachelor, and a leper. Leprosy is not exactly what I have, but what in the Bible is called leprosy (see Leviticus 13, Exodus 4:6, Luke 5:12–13) was probably this thing, which has a twisty Greek name it pains me to write. The form of the disease is as follows: spots, plaques, and avalanches of excess skin, manufactured by the dermis through some trifling but persistent error in its metabolic instructions, expand and slowly migrate across the body like lichen on a tombstone. I am silvery, scaly. Puddles of flakes form wherever I rest my flesh. Each morning I vacuum my bed. My torture is skin deep: there is no pain, not even itching; we lepers live a long time, and are ironically healthy in other respects. Lusty, though we are loathsome to love. Keen-sighted, though we hate to look upon ourselves. The name of the disease, spiritually speaking, is Humiliation.

Nov. 1. The doctor whistles when I take off my clothes. "Quite a case." ... The floor of his office, I notice, is sprinkled with flakes. There are other lepers. At last, I am not alone.... As I drag my clothes on, a shower of silver falls to the floor. He calls it, professionally, "scale." I call it, inwardly, filth (Updike, 1976, p. 28).

Baron (1985, p. 610) argues that any dermatologist who reads this description seriously thereby profoundly enriches his or her comprehension of the phenomenon of skin disease. The lived experience ("torture," "loathsome," "Humiliation") is not captured in the naturalistic account of the disease state ("psoriasis," "leprosy," "some trifling but persistent error in metabolic instructions"). Furthermore, says Baron, "there is a level on which it does not even matter whether the disease is psoriasis or leprosy; on such a level, one can talk seriously of 'Humiliation,' even as one decides whether to prescribe dapsone or psoralens with ultraviolet A."

In providing lifeworld descriptions of illness, clinical narratives expand upon (rather than supplant) the purely naturalistic account of disease states. Aleksandr Luria's books, *The Mind of a Mnemonist* (1987) and *The Man With a Shattered World* (1972), for example, give information not only about neuroanatomy and cerebral function, but provide profound insights into the lived experience of disorder which is produced by pathologies of memory and severe neurological deficit. In moving from pure medical description to a lifeworld account of neurological disorder, Sacks (1983, 1985c) focuses on the meaning that neurological disease has for his particular patients – and thus he gives an account of the chaotic disruption of body, self and world that such disorders necessarily entail. In his autobiographical account of a leg injury, Sacks (1984) details not only the specific neurological "fact" of his injury but recounts the experience of disordered body image which was an integral part of his illness – an experience which is shared by many patients with such injuries. In particular, Sacks' account reveals the profoundly distressing effect that such disturbances of body image create.

Furthermore, literary autobiographical accounts can provide a window into the experience of some illnesses where it is extremely difficult to gain firsthand descriptions through verbal communication with patients. For instance, Christopher Nolan's (1987) *Under the Eye of the Clock* is a remarkable account of what it is like to be severely handicapped with cerebral palsy. As John Carey notes with regard to Nolan's story:

[T]his is a voice coming from silence, and a silence that has, as Nolan is aware, lasted for centuries. He has a keen sense of the generations of mute, helpless cripples who have been "dashed, branded and treated as dross," for want of a voice to tell us what it feels like. Now that voice – or at any rate that redeeming link with a typewriter – has come, and we know. On page after page of this book, Nolan tells us. It should not be possible, after reading it, ever again to think as we have before about those who suffer what he suffers (Nolan, 1987, p. xii).

The profound gap between the immediate experience of illness and the naturalistic account of such experience is well illustrated in the clinical narratives provided by physicians who have themselves been sick. In describing their own illnesses most physicians quickly move beyond the traditional medical description of disease to a phenomenological description of the illness experience. In so doing they convey the disruption and disorder which characterizes illness-as-lived. In describing their experience they talk not of objective signs and clinical data but rather of loss of certainty (no longer feeling indispensable or indestructible). They recount fear and anxiety. They tell of losses engendered by the disruption of plans, goals and aspirations. They detail their isolation from others (particularly their sense of isolation from other physicians, including those who are treating them). They talk of losing their independence, of the indignities of becoming a patient. In particular, they focus upon the disruption of their lives which episodes of illness engender – disruptions that have a lasting impact and which do not necessarily cease if the disease is cured (Rabin, 1985a; Mandell and Spiro, 1987; Lear, 1980; Rosenbaum, 1988; Mullen, 1975; Sacks, 1984). Such narratives emphasize that illness is never encountered by the patient as an isolated entity but rather illness is always experienced within the context of a particular life and in light of personal hopes and aspirations.

The clinical narrative (the story of the illness as told by the patient) necessarily situates the illness within the larger context of the patient's life narrative. This is important in a number of respects. Earlier I noted that there is a distinction between suffering and clinical distress. In particular, I argued that suffering occurs at the reflective level and is experienced by persons, not merely by bodies. Consequently, suffering is intimately related to the meaning and significances assigned by the patient to immediate pre-reflective sensory experience. I further noted that, since suffering is intimately related to the patient's apprehension of illness, alleviation of suffering requires that explicit attention be paid to such apprehension. The clinical narrative provides important information with regard to the patient's biographical situation and, particularly, with regard to the meanings – both personal and cultural – which are a function of that biographical situation. As we have seen, such meanings determine the manner in which the patient construes illness and, furthermore, to a large extent such meanings determine whether "disease" involves suffering.

Moreover, it will be recalled that lived experience exhibits a certain temporal structure (i.e., that the temporal is constituted as a field of occurrence with past and future providing horizons for the present). Thus, it is important to recognize that illness in its immediacy is not simply an isolated physical event but rather it is an episode which is embedded in the unique life narrative of the patient. That is, the present "fact" of illness represents not so much an isolated instant along a given time-line as it does a present-now which must be considered against the horizons of past and future. In particular, present meaning is always constituted in terms of past meanings and future anticipations. Thus, the meaning of illness to a particular patient will depend upon the "collectivity of his or her meanings" – a collectivity which is necessarily a function of a particular life situation. In locating the illness within the context of a unique life narrative, the clinical narrative draws attention to the temporal structure of lived experience and, particularly, to the constitution of meaning which is a function of this temporal structure. In so doing, it provides at least an initial response to the question "What does this illness mean to you?"

*Summary*

In sum, then, the clinical narrative provides invaluable insights into the lived experience of illness. In reflecting the "voice of the lifeworld," rather than the "voice of medicine," the illness story relates the disruption of body, self and world (the disorder of lived body) which the immediate experience of illness represents. Furthermore, the narrative explicitly situates the illness within the context of a particular life and, in so doing, it discloses meanings inherent in that unique life situation – meanings which directly relate to the patient's experience of illness.

### 4. THE HEALING RELATIONSHIP

In his analysis of the manner in which a "communicative common environment" (a common world) is established between individuals Schutz (1962f, p. 318) notes that the "face-to-face" relationship is the predominant relation in the construction of the social world. In such a relationship the participants share time and space, perceiving one another. The participants share space in the sense that in this type of relation "the Other's body is within my actual reach and mine within his," and they share time in the sense that together they constitute what Schutz (1976a, pp. 171–172) has called a "vivid present." What characterizes the vivid present is a mutual experience of living simultaneously in several dimensions of time – that is, each participant in the relation not only experiences ongoing events in terms of a shared outer time but in the communicative process there is a synchronicity between the participants' ongoing flow of consciousness in inner time (in the sense that they are mutually directed to and engaged in experiencing an object or event in the world).[151]

Schutz (1962f, p. 317) notes that only in the "face-to-face" relation do we experience one another in our individual uniqueness. While the "face-to-face" relationship lasts we are "mutually involved in one another's biographical situation: we are growing older together." In this type of relation, then, the Other is perceived as a "co-subject" who has his or her own experiences of a "common" world.[152] It is in the context of the "face-to-face" relation that a communicative common environment may be established by means of which the participants attempt to construct a shared world of meaning.

The patient-physician relationship is a unique kind of "face-to-face" relationship in that the mutual involvement in one another's biographical situation (the shared world) is grounded in the patient's experience of illness. The patient comes to the physician because of some perceived lived body disruption (some unusual sensory experience or functional disturbance or some apprehension of an alteration in his or her body which is construed as "illness" or "disease"). And this perceived lived body disruption is the focus of the encounter.

Additionally, the patient-physician relationship has a specific end – the healing of the patient – and the relationship is entered into and perpetuated with this end in sight. The ill person comes to the physician seeking relief from lived body disruption, some means of alleviating or ameliorating suffering or distress.[153] In coming to the physician, says Pellegrino (1979a, p. 171), the patient does so with one specific purpose in mind: "to be healed, to be restored and made whole, i.e., to be relieved of some noxious element in his/her physical or emotional life which the patient defines as dis-ease – a distortion of his/her accustomed perception of what is a satisfactory life." The patient, then, is a suffering person who comes to the physician for assistance in regaining a former state of well being or, at least, a more optimal one.

Moreover, the sick person also comes to the physician seeking a means to communicate his or her dis-ease and thereby to make sense of this particular experience of illness. What the patient seeks is not simply a scientific explanation of the physical symptoms, but also some measure of understanding of the personal impact of the experience of lived body disruption. In communicating with the physician, the patient seeks to convey the meaning of illness in the context of a particular biographical situation.

It is important to note that the manner in which illness is conceived has a profound impact on the notion of "healing," and thus on the way in which the end of the patient-physician relationship is defined. If illness is attended to (and thus defined) exclusively in terms of "objective" pathophysiology (i.e., in terms of "disease states" which are manifested in demonstrated pathoanatomical and pathophysiological findings), the end of the medical encounter is understood to be primarily diagnosis and cure. The primary focus is on the disease state with a concurrent de-emphasis of the patient's lived experience of illness. If, however, illness is understood in terms of the primary experience of lived body disruption (and the concurrent disorder of body, self and world), then attention is focused on the dis-ease of the patient (and not solely the "disease" of the patient) and the goal becomes to restore the patient's integrity as a human being. This restoration of wholeness may include, but is not limited to, the restoration of bodily integrity.

This qualitative shift in emphasis (a shift which moves from a stance of confrontation with an abstract disease entity to a stance of addressing the existential needs of the person who is ill) has profound implications, in particular, for the response to those individuals facing chronic or incurable illness – where the restoration of health is not an attainable end. If the end of the medical encounter is defined solely in terms of diagnosis and cure of "disease," the suffering of those with chronic illness seems intractable. The focus on "cure" suggests that the physician has little to offer the person who is incurably ill.

As Baron (1981, p. 5) notes, such is the emphasis of the following statement appearing in *Harrison's Principles of Internal Medicine*, a textbook which is described as "probably the single most respected textbook of medicine."

The discovery and cure of potentially serious disease represents a far greater service to one's patients than ministrations in the course of an incurable condition (Thorn, 1977).

It is, of course, the case that (under optimal conditions) an important way to restore patients' autonomy and to return them to their former state of well-being is to cure the disease that impairs their autonomy (although it will be recalled that cure of disease does not always eliminate suffering) (Cassell, 1977, p. 18). However, the majority of patients who seek help from a physician are those suffering from illnesses which *cannot* be cured (Cassell, 1966, p. 149; Kleinman, 1988, p. 47; Leder, 1984, p. 35). For such patients the goal of cure, and the expectations which are derived from an acute disease model of illness, are inappropriate. For the chronically ill the control of disease is, by definition, limited. Rather the emphasis is necessarily upon reducing the disorder of body, self and world which the ongoing lived body disruption engenders. Indeed, Kleinman (1988, p. 229) argues that, in the case of chronic illness, the quest for cure is a "dangerous myth that serves patient and practitioner poorly." Such a "myth," he says, distracts attention from step-by-step behaviors that lessen suffering, even if they do not cure the disease.

With an emphasis on cure of disease the therapeutic focus is on likely medical interventions rather than on directly addressing the existential predicament of the patient. Little attention is paid, for example, to devising strategies to minimize such disturbances as the disruption of space and time, or to counteract the debilitating effects of fear and anxiety.[154] This emphasis on medical intervention and cure, in turn, arouses unrealistic expectations in the chronically ill patient. There is an overwhelming pressure "to do" something ("doing" being equated with instituting drug therapy or other medical treatment). Such pressure on both patient and physician alike may lead to inappropriate therapy. In addition, the inevitable failure of this therapy (inevitable in the sense that it cannot result in the complete restoration of health) leads to disappointment, frustration and helplessness.

As Engelhardt (1982, p. 41) has pointed out, the manner in which the disease state is construed has implications for how patients are treated and how they are regarded. An account that portrays diseases essentially as pathoanatomical and pathophysiological changes discounts patient complaints that are not easily referable to such changes. He notes that a "line is quite naturally drawn between bona fide complaints, those complaints accountable in pathoanatomical and pathophysiological terms, and male fide complaints, those complaints not amenable to such explanations." If illness is defined exclusively in terms of "objective" pathophysiology, then the goal of medicine is construed primarily as understanding and treating pathoanatomical lesions and pathophysiological disturbances. Consequently, the implication is that physicians have little to offer patients who suffer from complaints that do not fall into this category. Rather, as Engelhardt (1982, p. 53) indicates, such patients will be seen as "attempting to misuse medicine and to distract it from its important central and serious missions."

It is important to note, however, that many people who are ill do not have diseases which can be classified according to the conventional taxonomy (Engelhardt, 1982, p. 52; McWhinney, 1983, p. 5). Indeed, Ian McWhinney (1983, p. 5) argues that such illnesses account for at least half of the morbidity in the general population at large, and that they are capable of causing much chronic suffering and disability. As an example he cites studies of abdominal pain (in which only 21% of patients received a specific diagnosis after three months), headache (in which only 34.5% of 272 patients had received a specific diagnosis after one year and in which the remainder of complaints were classified as either "migraine" or "muscle contraction headache" – non-specific diagnoses in which he notes it was "virtually impossible to make a clear distinction between the two conditions"), and chest pain (in which only 50% of the patients studied received a specific diagnosis) (Watson, et al, 1981; Bass, et al, 1983; Blacklock, 1977).

If "cure" is perceived to be the goal, disease is the enemy and the patient's body the battlefield. The emphasis is on winning the war, whatever the cost. The "disease" is confronted as an abstract entity residing in, but in some way separated from the one who is ill.[155]

This emphasis on confrontation with a disease entity is reflected in the following description by Martin D. Netsky, a professor of medicine. He describes the treatment received by his dying mother in a large teaching hospital that prides itself on the excellence of patient care.

What happened was a nightmare of depersonalized institutionalization, of rote management presumably related to science and based on the team approach of subdivision of work.... Different nurses wandered in and out of my mother's room each hour, each shift, each day, calling for additional help over a two-way radio.... They were trained as part of a team "covering the floor" rather than aiding a sick human being.... Laboratory studies of blood and urine continued to be performed, fluids were given, oxygen was bubbled in, antibiotics were administered; the days went by but seemed to be years. The patient was seen occasionally by large groups of physicians making rounds, presumably learning the art of practicing medicine properly.... The chart was enlarged regularly with "progress notes." These hastily scrawled writings always dealt with laboratory data, never about the feelings of the patient or her family.... One report stated that occult blood had been found in the stool. Someone responded by writing in the chart that, in view of this finding sigmoidoscopic examination and a barium enema were indicated. I suggested to the author that his conditioned reflexive act was not warranted in the care of an unconscious 80-year old woman who wanted to die gracefully (Netsky, 1976, pp. 57–61).

Others have similarly noted that if the primary focus is the disease state, then the goal of medicine is considered to be the preservation of the body and biological life. Consequently, in an era of technological effectiveness life at all costs becomes not only a slogan but a reality (an imperative which may override other human values) (Cassell, 1977, p. 18).[156]

If cure is the overriding goal, inability to cure is equated with failure. Thus, the patient whose "disease" cannot be cured is often avoided as an uncomfortable reminder of failure. Dying patients who write of their illnesses relate their isolation from caregivers (Stoddard, 1978; Rabin, 1985b; Lear, 1980; Craig, 1991). As one terminally ill cancer patient noted, in the hospital "no one seemed to want to look at me" for to look at her "might have meant to see, in a place where only successful cure was acceptable, that she was incapable of being cured" (Stoddard, 1978, p. 21). The centering focus in biomedical science and medicine on curing disease and injury, says Zaner (1985, p. 240), suggests that those who cannot be "cured" not only "stand outside medicine as beyond its apparent powers, but also are living affronts to it. Being 'incurable' is being 'beyond help,' and this all too easily becomes the motive for being abandoned."[157]

If cure of "disease" is taken to be the overriding goal of the medical encounter, intractable illness poses a frustrating challenge to the physician's capabilities. However, if alleviation of dis-ease and suffering is perceived to be the end of the healing relationship, there is much the physician can do. Indeed, the doctor is perhaps one of the most effective allies that the patient can have in the struggle to deal with the limitations imposed by illness.[158]

Cassell (1966, p. 149) has distinguished between the "healing function" as opposed to the "curing function" of the physician. The "curing function" is, of course, limited to the cure of disease states. However, the "healing function" is directed at addressing and resolving the existential predicament of the person who is ill – at relieving (to the extent possible) the perceived lived body disruption which the illness engenders. Cassell notes that, in fact, in this day of cancer, chronic disease and the problems of the aging, the healing function of the physician is primary. As has been noted, patients with incurable disease far outnumber those with curable disease.

In the case of chronic or incurable illness the healing function of the physician is crucial. The healing function is not to be equated simply with giving reassurance, acceptance and patience (Cassell, 1966, p. 149). In a real way, through the healing relationship, the physician can restore the patient's integrity as a person. To do this the physician must address those factors which are fundamental to the experience of illness, such as loss of control, isolation, helplessness and loss of freedom to act. Whereas the restoration of wholeness may be limited in terms of restoring bodily integrity or eradicating "disease," the physician can assist the patient in regaining control (even if it is only limited control), overcoming helplessness and thus retaining the freedom to act. Although the freedom to act may be severely circumscribed by physical impairment, nevertheless the physician can and should assist the patient in continuing to live life to the fullest extent possible – to " live well" in spite of illness.

The healing function of the physician extends even to the dying patient. Physicians are only helpless in the battle against death if they see their role solely in terms of curing "disease." Once one recognizes one's function is to help the sick to the limit of one's ability, then one is almost always able to offer something, says Cassell (1966, p. 200). In the physician's care "the sick are protected from helplessness, fear, and loneliness, agonies that are worse than death."[159]

"Healing a person does not always mean curing a disease," argues Cicely Saunders, founder of St. Christopher's Hospice in London:

Sometimes healing means learning to care for others, finding new wholeness as a family – being reconciled. Or it can mean easing the pain of dying or allowing someone to die when the time comes. There is a difference between prolonging life and prolonging the act of dying until the patient lives a travesty of life (Stoddard, 1978, p. 175).

It may be objected that healing, so defined, is not limited to medicine. Pellegrino (1983, pp. 162–163) argues, however, that although psychologists, ministers, friends and families can provide healing relationships, they do so "over a limited range of human need." Persons who seek healing from a physician do so specifically because they regard themselves as sick, whereas persons who seek healing from those outside medicine do not consider themselves to be ill:

Sickness implies embodiment, the distinctly human phenomenon of a conscious self in a lived-body. When a person experiences some disturbance in his accustomed state of balance between body, psyche and self he counts himself as sick.

It is the fact of embodiment that creates the need for the physician. Only he can unravel the connections between the subjective experience of illness and its linkage to bodily function. Without denying the part others may play, the physician comes closest to what healing means – to restore wholeness or, if this is not possible, to assist in striking some new balance between what the body imposes and the self aspires to (Pellegrino, 1983, pp. 162–163).

Moreover, unlike family or friends, the physician is not personally involved in the disordered existence of the patient. Thus, the physician can act as the patient's confidant, advisor and arbiter, enabling the patient to clarify his or her situation and make appropriate decisions and adjustments.

The special relationship between physician and patient, the healing relationship, distinguishes clinical medicine from biomedical science, *per se*. In the healing relationship attention is focused on the experience of the one who is ill, rather than simply on the disease process itself. As Stephen Toulmin (1976, pp. 46–47) notes, in the traditional role of healer the physician's understanding is "typically particular rather than general, individual rather than collective, even (so far as is practicable) empathic rather than intuitive. He will focus attention entirely on the particular problems of individual patients, whatever these turn out to be, rather than view patients merely as 'nice cases of x-itis.'"[160] It is the explicit focusing on the particular problems of the individual patient that enables the physician to fulfill the role as healer. Healing includes the relief of suffering, of dis-ease, as well as the cure of "disease."[161]

Since healing involves the relief of suffering and the amelioration of the lived body disruption which illness engenders, it is clear that healing requires an understanding of illness-as-lived. The phenomenological analysis has revealed that suffering is always personal. It relates explicitly to the particular patient's life situation and to the meaning and significances which he or she attributes to the experience of illness. Thus, suffering may only be relieved if explicit attention is paid to the meaning that illness has for a particular patient within the context of a unique life world.

Moreover, the phenomenological analysis of body indicates that suffering relates not only to the loss of intactness of the biological body but to the loss of integrity of the whole web of interrelationships of body, self and world; that is, to the manner in which the patient *uniquely* exists his or her body and to the disruption of that embodiment which alters all relations and interactions with the surrounding world. If suffering is to be relieved, it is imperative directly to address this disruption of embodiment. This is particularly important in the case of chronic or incurable illness where it is not possible to restore the intactness of the biological body. The emphasis must be on addressing and ameliorating such problems as the disruption of lived spatiality and lived temporality, thus enabling the chronically ill patient to confront the ongoing disorder in an optimal fashion.

It has been shown that the disruption of embodiment includes not only the loss of bodily integrity but that it may result in a concurrent disruption of self (loss of integrity of the person). Cassell (1991) has noted that an important goal in the healing relationship should be maintaining the integrity of the person. Indeed, he suggests that suffering is "that state of distress induced when the intactness of the person is threatened or destroyed; such distress continues until the threat is gone or the integrity of the person is restored." In the case of chronic illness, Cassell (1990) notes that suffering may arise because the integrity of the person is threatened by internally generated conflicts (such as conflicts generated by the desire to fulfill the expectations of the social world, loss of self esteem, negative perceptions of the attitudes of others towards one's disability, and so forth). Knowing the patient and his or her values can allow the physician to assist the person with chronic illness to address such threats to the self. Even in the event that an illness is terminal, the integrity of self can be preserved. That is, mutual decisions can be made by physician and patient which preserve the patient's autonomy in the face of death (Cassell, 1977, p. 19).

## Summary

In sum, then, the physician-patient relationship is a unique kind of "face-to-face" relationship in that it is grounded in the patient's experience of illness and the relationship is entered into with a specific end in mind (the healing of the patient). The act of healing may include, but is not limited to, the cure of disease. However, it is evident that healing presupposes some understanding on the part of the healer of the patient's existential predicament. Such an understanding can only be reached if the physician (the healer) explicitly focuses on the illness as it is experienced by a particular patient. The act of focusing on the patient's lived experience requires that the physician temporarily set aside the naturalistic interpretation of illness as a disease state in order to attend to the lifeworld disruption which illness engenders.

In this respect it is important to note that, as healer, the physician functions not only as scientist but as colleague in the physician-patient relationship. An adequate understanding of the patient's lived experience is important both for the physician acknowledging the patient as person (i.e., in his or her role as colleague) and for the physician treating the patient *qua* scientist (i.e., the physician must have an adequate understanding of the lived experience in order to bring to bear his or her scientific knowledge in devising effective therapy for the patient).

# NOTES

[1] It should be noted that in Edmund Husserl's work there is an important distinction between psychological phenomenology and transcendental phenomenology. Both involve the phenomenological reduction but they do so at different levels. The first level (that of psychological phenomenology) involves the "suspension of belief" in the "outer world" in order to make explicit the field of consciousness – the "noetic" (the intentional act which presents the thing as meant), and the "noematic" (the thing as meant). The transcendental level involves a further reduction which "brackets out" not only the "outer world" but also individual consciousness, in an attempt to disclose the ultimate structure of consciousness. Throughout this work I shall be carrying out psychological phenomenological analysis and not transcendental phenomenological analysis. Furthermore, I do not intend to confine my analysis to the work of any one phenomenologist. Rather I intend to show that the work of several phenomenologists provides genuine insights into the different perspectives of physician and patient. At various points in the text I will identify important differences between these thinkers, as such differences bear on the problem under discussion.

[2] As I have noted above, I am confining my analysis to the level of psychological, rather than transcendental, phenomenology. Thus, wherever I refer to "phenomenological" description or a "phenomenological" approach in this work, I am intending this to refer to psychological phenomenology.

[3] For a concise discussion regarding intentionality see Edmund Husserl's (1929, pp. 699–702) article on phenomenology in *Encyclopaedia Britannica*. As Husserl notes:

In unreflective consciousness we are "directed" upon objects, we "intend" them; and reflection reveals this to be an immanent process characteristic of all experience, though infinitely varied in form. To be conscious of something is no empty having of that something in consciousness. Each phenomenon has its own intentional structure, which analysis shows to be an ever-widening system of individually intentional and intentionally related components.... Phenomenological psychology's comprehensive task is the systematic examination of the types and forms of intentional experience ...

[4] Merleau-Ponty (1964) insists that the field of ideality is necessary in order to give an account of meaning. In other words, we must step back from our ongoing involvement in the world in order to render the meaning of that involvement explicit. The purpose of phenomenological reflection is to grasp the *sense* of experience rather than simply living through it.

[5] As Husserl (1962, p. 100) notes, the phenomenological reduction prevents one from using "any judgment that concerns spatio-temporal existence."

Thus all sciences which relate to this natural world, though they stand never so firm to

121

me, though they fill me with wondering admiration, though I am far from any thought of objecting to them in the least degree, I disconnect them all, I make absolutely no use of their standards, I do not appropriate a single one of the propositions that enter into their systems, even though their evidential value is perfect, I take none of them, no one of them serves me for a foundation – so long, that is, as it is understood, in the way these sciences themselves understand it, as a truth concerning the realities of this world. I may accept it only after I have placed it in the bracket. That means: only in the modified consciousness of the judgment as it appears in disconnexion, and not as it figures within the science as its proposition, a proposition which claims to be valid and whose validity I recognize and make use of (Husserl, 1962, p. 100).

It should be noted that the phenomenological commitment to radical reflection distinguishes phenomenology from other philosophical approaches which appeal to "experience" (for example, those that begin with such constructs as "sense data," etc.) For a further discussion of this point see Kohak (1978, pp. 152–161).

6 Merleau-Ponty argues that, because we are "through and through compounded of relationships with the world," the only way for us to become aware of that fact is to put such relationships "out of play" and make them explicit. The reduction, he says, discloses the "facticity" of the world and brings to light our inescapable embeddedness in the world (our being-in-the-world or our being-towards-the-world). It should be noted that, in this respect, Merleau-Ponty differs from Husserl (1962) in that the latter argues that the "bracketing of the world" ultimately discloses the "world" as the intentional correlate of transcendental subjectivity. As noted above, in employing the reduction, Husserl states that there are different levels of reduction – the ultimate being the transcendental level (the realm of pure consciousness) – and, in order to address the philosophical problems of epistemology, he finds it necessary to take everything back to the transcendental level. However, I would argue with Merleau-Ponty and Schutz that, whereas the transcendental phenomenological reduction is necessary in order to make explicit the intentional structures which determine meaning, it is possible to provide a rigorous psychological phenomenological description of what is given without taking everything back to the transcendental level.

7 Merleau-Ponty notes that in our experience we can distinguish the *fact that* we are living through something from *what it is* that we are living through in this fact. Kohak (1978, pp. 13–22) makes this same point when he argues that Husserl provides the insight that we always experience particular objects as embodying a "principle" which we can grasp separately. That is, we can not only reflect on but perceive our experience from both "a factual and an eidetic viewpoint, from the standpoint of preoccupation with particulars as well as 'in principle,' focusing our awareness on the principle an instance embodies." Lived experiences are intelligible in that they exhibit a typical way of being (in Kohak's terms, "demonstrate a principle"). For instance, the experience of being-afraid has a "logic of its own" which can be grasped apart from a particular instance of an individual's fearing. Insofar as I am able to grasp an intelligible structure that imposes itself on me whenever I think of an intentional object, then I go beyond the contingent fact and arrive at an insight into its invariant features (Merleau-Ponty, 1964, p. 55).

[8] In order to arrive at the invariant features of phenomena, phenomenologists employ the method of "imaginative free variation (Husserl, 1962, pp. 181 ff). Free phantasy variation differs from empirical generalization in that, in the former, one explicitly endeavors to consider a range of actual and possible affairs *as examples* of some kind or sort in order to determine which characteristics intrinsic to the range of variations are invariant. As Zaner (1981, p. 193) notes, "it is never a matter of trying to generalize ... it is rather a question of trying to determine what is invariantly common to (exemplified by) every actual and possible example of the kind in question." For a helpful discussion on free phantasy variation, see Zaner (1973a and 1973b) and Bachelard (1968, pp. 173–197). As Zaner (1973a and 1973b) notes, the key point is that the individual affair is taken *as exemplifying* some kind or sort and the phenomenologist's concern is for what-is-exemplified by that, or any other possible example. The concern is with types rather than tokens (with, for example, the triangle as such and not the illustration of a triangle in the textbook). See also Natanson (1973, pp. 67 ff.). For a discussion of the value of free-phantasy variation as a method for disclosing the invariant see, Zaner (1981, pp. 242–249).

[9] This is not to say, however, that such an analysis may not have implications for empirical psychology. In particular, the psychological phenomenological analysis provides the means to clarify the concepts used by empirical psychology.

[10] All empirical sciences begin with the presupposition of the lifeworld as already given and amenable to their methods and theories. Therefore, they presuppose the kind of thing that phenomenology tries to elucidate – namely, the meaning structures through which our "coming to know" objects in the world is first of all made possible.

[11] For a helpful discussion of the methodological distinction between the empirical-positivistic tradition and the phenomenological approach see Natanson (1969, pp. 85–110). Psychology is a positive science not a reflective discipline. As an empirical science psychology does not reflect critically on the origins or sources of the claims that it makes. Furthermore, empirical psychology involves a physicalistic interpretation of consciousness, whereas psychological phenomenology is interested in describing the structure of consciousness, its essential character as the basis of perceptual reality.

[12] In this respect I should note that the phenomenological exploration of intersubjectivity may be carried on at different levels. In *Cartesian Meditations* Husserl (1982, pp. 89 ff.) seeks to trace the constitution of an intersubjective world to the ground of the transcendental ego. However, at another level (the level of phenomenological psychology) the task is to provide a descriptive account of the essential features of the world taken as intersubjective in the natural attitude and that is the focus of this work.

[13] It should be emphasized that the only way to make explicit the nature of this distinction is to focus upon the manner in which meaning is constituted differently by physician and patient, and this requires that one "set aside" theoretical commitments derived from the natural sciences in order to describe what gives itself directly to consciousness.

[14] Once again, in order to answer the question "What is the body experienced as?", it is imperative to conduct such an inquiry without presupposing that the natural scientific account is the only legitimate account. In order to do that, it is necessary to engage in a psychological phenomenological reflection which does not presuppose the

orientation of natural science towards the reality of the object or events with which it deals. Furthermore, phenomenologists are committed to the position that not all meaningful questions are in principle capable of being answered by the methodology of the positive sciences.

[15] It should perhaps be noted that phenomenological approaches to the problem of illness are not altogether new. All such analyses emphasize that illness is intelligible as a lived experience – an experience which can be rigorously examined and elucidated. However, the psychological phenomenological analysis carried out in the present work aims to contribute in a more particular way in that it directly focuses on and clarifies the different perspectives of physician and patient which are often in conflict. For a work which takes an explicitly phenomenological approach to the problem of illness see Kestenbaum (1982b).

[16] Throughout this work I shall use Gerhard Bosch's (1970) designation of "own world" to refer to the private, egoistical world of the individual (the world of originary experiencing) as opposed to the "common world" (the intersubjective world in which understanding with others has been established and about which one can communicate). It is in the exploration of "own world" and "common world," in particular, that psychological phenomenology can provide insights into the manner in which meaning is constituted differently by physician and patient.

[17] In this chapter I have focused on the psychological phenomenology of Edmund Husserl and Alfred Schutz since I find their work to be especially helpful in elucidating meaning within the worlds of physician and patient. It should perhaps be noted that, while Husserl and Schutz both provide an analysis of the eidetic structure of experiencing, they differ in some important respects. Schutz's main concern is to analyze the meaning structure of the intersubjective world of everyday life (i.e., to provide a descriptive phenomenology of the "natural attitude"), and he finds Husserl's analysis of transcendental intersubjectivity problematic.

[18] This naive, unreflective experiencing of the world is what Husserl has called "the general thesis of the natural standpoint" or the "natural attitude." In a familiar passage Husserl says:

I find continually present and standing over against me the one spatio-temporal fact-world to which I myself belong, as do all other men found in it and related in the same way to it. This "fact-world," as the word already tells us, I find to be *out there*, and also *take it just as it gives itself to me as something that exists out there*. All doubting and rejecting of the data of the natural world leaves standing the *general thesis of the natural standpoint*. "The" world is as fact-world always there; at the most it is at odd points "other" than I supposed, this or that under such names as "illusion," "hallucination," and the like, must be struck *out of it*, so to speak; but the "it" remains ever, in the sense of the general thesis, a world that has its being out there (Husserl, 1962, p. 96).

I shall discuss the "natural attitude" in further detail later in this chapter.

[19] For an excellent commentary on Husserl's investigation of time, see Sokolowski (1974).

[20] Schutz has used this term to refer to the unique biography of each individual

(Natanson, 1962, p. xxix).

[21] For a further elaboration of this point see Natanson (1962, p. xxviii ff.).

[22] Schutz (1962c) describes the process of typification as follows:

The factual world of our experience ... is experienced from the outset as a typical one. Objects are experienced as trees, animals, and the like, and more specifically as oaks, firs, maples, or rattlesnakes, sparrows, dogs. This table I am now perceiving is characterized as something foreknown and, nevertheless, novel. What is newly experienced is already known in the sense that it recalls similar or equal things formerly perceived. But what has been grasped once in its typicality carries with it a horizon of possible experience with corresponding references to familiarity, that is, a series of typical characteristics still not actually experienced. If we see a dog, that is, if we recognize an object as a dog, we anticipate a certain behavior on the part of this dog, a typical (not individual) way of eating, of running, of playing, of jumping, and so on.... In other words, what has been experienced in the actual perception of one object is apperceptively transferred to any other similar object, perceived merely as to its type (Schutz, 1962c, pp. 281–282).

Schutz notes that such "typifications" comprise the individual's stock of knowledge by means of which he or she is able to interpret the totality of everyday experience. The "typifications" which comprise the individual's knowledge of the world (what Schutz has termed his "stock of knowledge-at-hand") are derived either from one's own previous experiences, or are handed down by others such as parents or teachers. From childhood on the individual continues to add to the stockpile of "typifications." Thus, the world of everyday life assumes a familiar quality which makes the prediction and control of experience possible.

[23] In this connection Schutz (1962f, pp. 308–309) argues that the world is organized around the individual in terms of (1) the "world within actual reach" (the sector of the world which I can modify directly by movements of my body or with the help of artificial extensions such as tools); (2) the "world within potential reach" (the world of my potential working acts); and (3) the "world within restorable reach" (my recollections of the world within my reach in the past).

[24] Jerome Bruner (1987, pp. 11–32) argues that we have no way of describing "lived time" save in the form of a narrative. This is not to say that other temporal forms cannot be imposed on the experience of time but none of them, he says, succeeds in capturing the sense of *lived* time, of "living through" our experiences.

[25] Husserl (1970b, pp. 379–383) has termed the world of everyday experience the "lifeworld" in order to distinguish it from, for example, the world of science.

[26] This, of course, is the crux of the problem of intersubjectivity. How (if at all) is it possible to account for intersubjective agreement, shared meaning, a "common" relationship to an "objective" world given the unique nature of experience? Husserl (1982, pp. 89–151) addresses the problem of the constitution of the Other and the constitution of the objective world in the *Fifth Meditation*. However, it should be noted that his account of transcendental intersubjectivity is not entirely satisfactory. For critical comments of this account see Schutz (1975).

[27] This consciousness of being both subject and object is rendered explicit in the

experience of illness and in the doctor-patient encounter. In the clinical encounter the body is objectified in the sense that it is attended to as an exclusively biophysiological mechanism and as an object for scientific investigation. In the experience of being looked-at, under the "gaze" of the physician, the patient concretely recognizes the duality of his or her being-an-object for the Other and concurrently his or her being as a suffering subject.

[28] As noted above, Schutz finds Husserl's account of transcendental intersubjectivity problematic and he argues that intersubjectivitiy is an irreducible fact of the lifeworld.

[29] Schutz (1962f, pp. 313–315) follows Husserl in arguing that while the Other's body is given to me as an originary presence, his or her psychological life is not presented but rather appresented.

[30] In explicating the constitution of intersubjectivity phenomenologically, Husserl (1982) has noted that there is a distinction between my primordial sphere of originary experiencing and the "merely presentiated primordial sphere" of the Other. Consequently, there is a distinction between two noetic strata each of which can be explicated but neither of which can be identified with the other.

[31] As Bosch (1970, p. 55) notes, it is through communication that the individual erects a bridge between the world that he or she essentially experiences as "own world" and the world of the Other. This communication is based on original experiences which "so long as they are not revealed by the act of communicating, remain a concrete part of the own world, and even when they are communicated they still retain an abstractable portion of own-world originality."

[32] For a helpful discussion on this shift in attention see Zaner (1973a).

[33] Husserl (1970a) has noted the distinction between the lifeworld (the world of everyday experience) and the world of science in terms of both the distinction between the "natural" and "naturalistic" attitude, which is explicated in the next section, and in terms of the difference between the goals of the lifeworld and the goals of the world of science. The goal of science is to determine nature "in-itself" through "truths in themselves." Thus, the world of science is first and foremost a purposeful structure which is constituted within a horizon of already existing scientific works. The lifeworld on the other hand is constantly pregiven, valid constantly and in advance as existing, but not valid because of some purposeful investigation. Thus, although the lifeworld represents a "structure" it is nevertheless not a "purposeful structure" in the way that science is. Rather, the lifeworld "was always and continues to be 'of its own accord.'" Consequently, the world of science constitutes a domain within and presupposes the lifeworld. Patrick Heelan (1973) argues that a crisp demarcation between the manner in which entities are experienced in the lifeworld as opposed to the world of science cannot be maintained in the case of certain kinds of scientific activity. In particular, he suggests that the observable scientific entities of experimental science belong to the lifeworld in that they are experienced by the practicing scientist in an originary manner. It is important to note, however, that such scientific observations are preceded by many complex inferences and training (e.g., the untrained individual using a microscope cannot make observations of mitochondria and Golgi bodies because facets of these entities have to be learned). In other words, the scientific "habit of mind" (which incorporates a complex stock of acquired knowledge, a certain noetic intention which animates the inquiry, and so

forth) determines the manner in which "reality" is constituted.

³⁴ In the following chapter I will discuss in detail the manner in which illness is constituted by the patient. As I shall note, it is the case that the patient's experiencing of illness is influenced by the theoretical understandings that are embedded in the lifeworld. Nevertheless, illness-as-lived is fundamentally experienced in terms of the disruption of body, self and world. Consequently, this lived body disruption is the focus of the patient's concern.

³⁵ The prevailing model of illness (the biomedical model) will be explored in detail in the next chapter.

³⁶ The constitution of the disease state will be explored at length in Chapter Two.

³⁷ For a detailed account of the temporality of illness see Toombs, "The Temporality of Illness," (1990).

³⁸ In relating his or her experience of illness to the physician, the patient must do so in terms of "objective time" (since "objective" time is the common language for time). However, the patient experiences the illness in "lived" time.

³⁹ While I agree with Schutz that the individual organizes the world into strata of major and minor relevance according to his or her unique life plan, it is not clear to me that the recognition of "fundamental anxiety" is explicit. Although at some level each of us is aware of our own mortality, for the most part we do not – in any deep sense – consciously reflect upon this mortality. Perhaps Schutz's point is just that many of the relevances that pertain to our life situations are pragmatic relevances basic to perpetuation of life (e.g., I go to work to earn money to provide food and shelter for myself and my family). Heidegger (1962, pp. 296-299) argues that much of the individual's time is spent in inauthentically denying the "fundamental anxiety."

⁴⁰ Engelhardt notes that since the expertise of physicians tends to be non-global, physicians narrow down their focus of concern still further to what they can alter or change. Thus, surgeons think of surgical interventions, internists think of medical interventions, psychiatrists think of psychiatric interventions, etc. The result is that physicians focus upon only bits and pieces of the whole that is lived through by the patient.

⁴¹ In this respect Schutz (1976a) notes, for example, the distinction between experiencing a piece of music and constituting its meaning in terms of the ongoing flow of consciousness in internal time, and grasping a piece of music as a typical example of, say, "a sonata." In the first case the meaning is constituted polythetically (i.e., as a series of steps in inner time), in the latter it is constituted monothetically (i.e., without reference to the polythetic steps in which the music is experienced in its particular individuality). See also Husserl (1962, pp. 307–311).

⁴² Obviously, here it would seem that I part company with Schutz in that I suspect that the explicit awareness of "fundamental anxiety" only makes itself concretely felt in moments of existential crisis.

⁴³ For an interesting analogy to the patient's experience see Schutz (1976b, pp. 106–119). Schutz notes that the person who returns home after spending a period of time away in a different environment finds himself unable to communicate his experience to those who have remained at home. He is unable to do so because he can no longer communicate on the basis of a shared set of typifications. As an example, Schutz quotes the case of the returning veteran:

When the soldier returns and starts to speak ... he is bewildered to see that his listeners, even the sympathetic ones, do not understand the uniqueness of these individual experiences which have rendered him another man. They try to find familiar traits in what he reports by subsuming it under THEIR preformed types of the soldier's life at the front. To them there are only small details in which his recital deviates from what every homecomer has told and what they have read in magazines and seen in the movies. So it may happen that many acts which seem to the people at home to be the highest expression of courage are to the soldier in battle merely the struggle for survival or the fulfillment of a duty, whereas many instances of real endurance, sacrifice and heroism remain unnoticed or unappreciated by the people at home (Schutz, 1976b, pp. 106–119).

[44] For an interesting discussion on the role of typification in the diagnostic process see Schwartz and Wiggins (1987).

[45] The complexity of specialized knowledge is such that its acquisition is, in principle, no longer accessible to all. Thus, there arises the need for not only "specialists" but "sub-specialists" each of whom has the requisite knowledge of a small domain of specialized knowledge. As specialized knowledge becomes more complex and differentiated, the various provinces of special knowledge become progressively further "removed" from general knowledge and the gap between "laymen" and "experts" becomes greater (Schutz and Luckmann 1973, pp. 314 ff.).

[46] As was noted above (see note 16) the phenomenological analysis of intersubjectivity has shown that differences in individual perspective can never be fully overcome given the nature of consciousness. Schutz has argued that common sense thinking overcomes differences in individual perspective by means of "the *idealization* of the interchangeability of the standpoints." I am arguing that this idealization fails in the case of inner events such as illness.

[47] Schutz (1962f) argues that one of the essential ways in which the Other's world transcends mine is that we each experience events uniquely in inner (rather than objective or outer) time. Consequently, the establishment of a "communicative common environment" presupposes individuals sharing an experience of an event in outer time. Schutz comments, "I and you, WE see the flying bird. And this occurrence of the bird's flight as an event in outer (public) time is simultaneous with our perceiving it, which is an event in our inner (private) time. The two fluxes of inner time, yours and mine, become synchronous with the event in outer time (bird's flight) and therewith one with the other" (Schutz, 1962f, p. 317).

[48] I shall have more to say about the manner in which the meaning of illness is constituted at various levels by the patient, beginning with pre-reflective experience, in the next chapter.

[49] In this respect it is worth noting again that what is of primary relevance to the patient is life disruption and that such disruption may be engendered not only by the illness but also by the treatment. Consequently, symptoms which may be considered merely as "side effects" of treatment by the physician (and consequently regarded by the physician as perfectly acceptable in the treatment plan) may be experienced by the patient as unacceptable in that they cause a significant disruption of his or her life plan. A recent experience with monthly chemotherapy treatments to slow the

progression of multiple sclerosis has illustrated this for me in a personal way. The symptoms of nausea, vomiting, weakness, and loss of appetite which accompany chemotherapy are, for the most part, not considered to be a serious problem by the physician but they are profoundly disturbing to the patient who experiences them.

[50] A portion of the material in this chapter has been published in Toombs (1990), "The Temporality of Illness: Four Levels of Experience."

[51] In this context I am focusing for the most part on illness which relates directly to the disruption of body rather than illness which is experienced as a disorder of mental functioning. This is not intended to suggest an arbitrary distinction, or to imply that a careful analysis of mental illness is not warranted.

[52] It should perhaps be noted that such unusual sensory experience does not have to be pathological for this shift of attention to occur. One may, for example, become acutely aware of one's body in moments of sexual arousal, during strenuous physical exercise, and so forth. The key point is that the shift of attention renders the body itself thematic.

[53] In the next chapter I shall distinguish between the patient's perception of his or her body as a physiological organism and the physician's perception of the patient's body as a scientific object. I shall note that, in perceiving his or her own body as a physiological organism, the patient experiences the body as "uncanny." The experience of the body as "uncanny" is quite different from the conception of the body as a scientific object. Similarly, in grasping the lump in one's breast as "cancer," one conceives of one's illness in largely metaphoric terms (having to do with the existential implications of the illness), as opposed to grasping it as the "disease state" which is known by the physician.

[54] In this connection Engelhardt (1989, pp. 71–79) says that pre-linguistic sensations must necessarily be described linguistically in the medical encounter and that language itself interprets and shapes the experience.

[55] Schutz makes this point when he argues that a completely uniform social distribution of knowledge cannot exist in principle because there are different social relevances based on gender, age, social strata, and so forth (Schutz and Luckmann, 1973, pp. 310 ff.).

[56] As Engelhardt has pointed out, in a highly technological society such as ours "bits of knowledge" acquired from others are incorporated into the lived experience of the body. Consequently, the boundaries between "suffered illness" and "disease" may not be as crisp as Sartre suggests, in that the apprehension of "suffered illness" as a synthetic totality will be influenced by the theoretical understandings that are embedded in the lifeworld. The apprehension of "suffered illness" will reflect the particular lifeworld in which the patient finds himself.

[57] It should be noted that in this context I am confining my analysis to western scientific medicine. Kleinman and others note, of course, that illness is conceptualized differently by practitioners in different systems of medical knowledge (Kleinman and Mendelsohn, 1978, pp. 314–330; Cassell, 1976, pp. 27–37). Furthermore, the actual construal of the "disease state" has varied in different times. See Engelhardt (1982, pp. 41–57). Nevertheless, as explanatory models, these differing conceptualizations share in common the fact that they represent a level of constitution which is distinct from the level of lived experience.

[58] It should, of course, be noted that the constitution of the disease state has changed through time. For an excellent summary of the changes in understanding which have led to this modern pathoanatomical and pathophysiological account see, Engelhardt (1982). Engelhardt notes, for example, that there has been a shift from clinically oriented appreciations of disease – such as that of Thomas Sydenham (1624–1689) – to pathoanatomical, pathophysiological and bacteriological accounts of disease. Consequently, the theoretical presuppositions of the basic sciences have come to structure the experience of illness. He argues that, largely as a result of developments in the science of pathoanatomy in the 19th century, the primary focus of medicine went inside the body and disease thus became identified with pathoanatomical lesions or pathophysiological disturbances. See also the following works: Sydenham (1981, pp. 145–155); Morgagni (1981, pp. 157–165); Bichat (1981, pp. 167–173); Virchow (1981, pp. 187–195); Cohen (1981, pp. 209–219); Temkin (1981, pp. 247–263). This shift between 18th and 19th century understandings of disease is also detailed by Foucault (1975).

[59] That this account of illness is problematic is well illustrated by Baron (1985). In this article he refers to several studies on ulcer treatment which all demonstrated that the anatomic fact of an ulcer cannot be correlated with patients' complaints. For example, in one major study after four weeks of treatment 55 percent of patients whose ulcers were unhealed were asymptomatic, regardless of treatment group. In addition 12 of 45 patients whose ulcer had healed endoscopically continued to have ulcer symptoms. See Peterson, *et al* (1977); Lauritsen, *et al* (1985).

[60] It can be argued, of course, that all facts (including medical facts) are interpreted facts. Ludwik Fleck has demonstrated that medical facts are culture-laden. That is, there is an intimate relationship between the observer and the observed such that observations are always culture-laden. Certain thought-styles make certain observations possible and others impossible. See Trenn (1981, pp. 237–256); McCullough (1981, pp. 257–261). For the relation between facts and theories see, Kuhn (1970).

[61] Cassell (1986) has suggested that one's life is a project or process such that the life project may be seen as a "fabric wherein a pattern is being created of each moment." Thus, it is useful to see illness as, among other things, a disruption of this pattern.

[62] For an enlightening discussion on the social aspects of disability see Murphy (1987). Murphy, a distinguished anthropologist at Columbia University, became quadriplegic as a result of a spinal cord tumor. In describing his experience he pays special attention to the social meanings which accompany disability. Herzlich and Pierret (1987) make the important point that in a society in which we define ourselves as producers, the incapacity to perform takes on a particular significance. Indeed they argue that reduced ability to perform and the resulting enforced inactivity have today become the essential perception of the sick body. I would also argue that the emphasis on performance in our society is a direct source of suffering for the disabled in that disability almost always affects one's capacity to produce causing a diminishment in self worth and social status. Concrete suffering is also caused, of course, by loss of employment and educational opportunities. For an excellent discussion on the effect of disability on women, in particular, see Asch and Fine (1988). The authors note that while the views of the non-disabled (both women and men) towards persons with disabilities are overwhelmingly negative, studies suggest that the disabled woman

may be more negatively viewed than the disabled man in this culture. For instance, Hanna and Rogovsky (1986) asked non-disabled college students to draw associations for "woman" and "disabled woman." "Woman" drew positive associations of worker, of sexuality, of mother or wife whereas "disabled woman" drew associations of dependence and impairment (crippled, almost lifeless), of age (gray, old, white hair), of despair (someone to feel sorry for, pity, lonely, ugly). The disabled woman was virtually never depicted as wife, mother or worker by the more than one hundred students interviewed. When asked how women and men using wheelchairs became disabled, students attributed male disability to external situations such as war, work injury or accident. They attributed female disability to internal causes such as disease. The authors suggest that attributing disability to disease may foster more negative attitudes (fear of contagion, moral culpability, and so forth).

[63] Engelhardt (1976, p. 262) argues that particular ideologies are likely to tempt us to explain particular phenomena as diseases in order to fit our ideological needs (e.g., drapetomania, the disease of slaves who fled the South for the North). Also he notes that states of affairs are classified as disease states for social and ideological reasons, to apply the sick role to those in that state (e.g., alcoholism, drug abuse, and so forth). For a fascinating discussion of the interrelationship between evaluation and explanation in this regard see Engelhardt (1974). In this article Engelhardt describes how in the 18th and 19th century masturbation was regarded to be a dangerous disease entity.

[64] As Frank notes with regard to his own experience of cancer:

Whenever I told someone I had cancer I felt myself tighten as I said it. Saying the word "cancer," my body began to defend itself. This did not happen when I told people I was having heart problems. A heart attack was simply bad news. But I never stopped thinking that cancer said something about my worth as a person. This difference between heart attack and cancer is stigma....

My heart attack damaged my body but did not stigmatize it.... During my heart problems I could no longer participate in certain activities; during cancer I felt I had no right to be among others. As much as I disliked being in the hospital, at least there I felt I belonged. I knew this was foolish. I didn't belong in the hospital; I was hiding there. Ill persons hide in many ways. Some begin to call cancer "c.a.," "the big C," or other euphemisms. I called it cancer, but as I said it I felt that tightness (Frank, 1991, p. 92).

[65] In writing of her experience Barbara Webster (1989) makes the point that the impact of a diagnosis such as M.S. is devastating emotionally, as well as physically. The knowledge that one has a potentially crippling and devastating disease necessarily affects every aspect of one's life including one's sense of self, one's relations with others, and one's conceptions about the future. This is the case, even in the event that one is experiencing mild physical symptoms. Webster recounts that her suffering would have been greatly alleviated if attention had been focused on the global impact of her diagnosis.

[66] Not only is it the case that the lived experience of illness may be present in the absence of demonstrable pathoanatomical or pathophysiological findings but the reverse is also true. That is, illness may be construed as a "disease state" in light of

certain objective clinical findings (abnormal x-rays, blood tests, etc. discovered during a routine physical exam) but the patient has noticed no abnormal sensory experience (i.e., he or she doesn't feel sick) and, therefore, the patient has not grasped his or her experience as illness at the level of immediate experience. Feinstein (1967, pp. 145–146) refers to such patients as "lanthanic," as opposed to "complainant" patients. A "lanthanic" patient is one whose disease is discovered "accidentally," i.e., he or she has not come to the physician because of symptoms relating to the disease which is discovered. An example of a "lanthanic" patient would be "an asymptomatic man who, during a 'routine check-up' is noted to have pathognomonic abnormalities in his electrocardiogram." Since he has noticed no abnormal sensory symptoms, he has not grasped his experience as illness or, more particularly, as "coronary artery disease."

[67] For a personal account of this type of suffering see Webster (1989, pp. 1–15). After several years of being told there was nothing physically wrong with her (years which caused her intense suffering because her experience of illness was not validated), Webster was finally diagnosed with multiple sclerosis.

My first reaction to receiving a diagnosis of multiple sclerosis was one of shock and terror, mixed with a deep sense of relief. I was stunned and I knew very little about the disease, which added to my terror, but at the same time there was overwhelming relief in knowing there was a solid physiological reason for the symptoms and inexplicable bouts of illness I had experienced over the years.

Receiving a diagnosis, any diagnosis, therefore, gave me a sense of great relief. I was not merely neurotic; there was a reason for what had happened to me over the years. *I felt immensely vindicated and affirmed* (Webster, 1989, p. 23). (Underlining mine).

[68] I do not believe that it is possible for patients ever to conceive of their illness exclusively in terms of pathophysiology since patients necessarily live their illnesses. That is, the patient's perspective never completely merges with the physician's – although here it is coming closer. For example, talking of her own experience with diabetes, the physician, Tamara Bell states:

In second-year pathophysiology, we were taught that in the diabetic patient, blood glucose was determined by three controllable factors: diet, exercise, and insulin.... So, I spent a month strictly controlling diet, exercise and insulin in an effort to achieve mastery over my blood glucose. In addition to having a miserable month, I found that the glucose levels remained somewhat spurious.... I had lost touch with what was so clear to me as a ten-year old – namely, that blood glucose is determined by countless factors, many of which have a life of their own (that is, ambient temperature, hormonal milieu, emotional state, etcetera).... So while Tami the doctor is concerned about a number on a glucometer, Tami the patient is concerned about how she has felt in the last six to eight hours.... While Tami the doctor insists on tight control *over* her disease, Tami the patient realizes she may have to settle for peaceful coexistence with her disease (Bell, 1991, pp. 29–30).)

[69] Earlier I commented that "objective" clinical data take on a life of their own. Not

only do such data have significance for the physician but they also affect the patient's perception of his or her illness. For example, although in my daily activities I have been aware of a progressive loss of function over the past several months, receiving an "objective" evaluation which stated that I had progressed from four and a half to seven and a half on the Kurtzke Disability Scale caused me to view this loss of function as even more disruptive and threatening than I judged it to be prior to the evaluation.

[70] In his analysis of temporality, Husserl (1964, pp. 57–59) distinguishes recollection (or secondary memory) from retention (primary memory). In retention the past temporal phases of an object are retained in primary memory as part of the present consciousness of the object whereas, in recollection, the object is no longer actually perceived but is re-presented in an "as-if" presentation.

[71] In this regard Cassell (1985a, p. 18) has noted that the causal story of an illness occurs at several different levels: macromolecular, involving say the platelets; the organ level, involving blood vessels and lungs; the whole organism level, the person's functioning as a whole organism, and so forth.

[72] A portion of the material contained in this chapter has appeared in S. Kay Toombs, "Illness and the Paradigm of Lived Body" (Toombs, 1988).

[73] In *Ideas* II (1989, p. 152) Husserl focuses on the body as the bearer of localized sensations and he makes the point that if I touch my left hand with my right, I experience the left hand as an object. Furthermore, he argues that in all experiencing of spatial objects the body as perceptual organ of the experiencing subject "is involved" (Section #36). There is a sense in which I have a non-thematic awareness of body at the pre-reflective level. However, to focus on the body *as* perceptual organ, or to consider the manner in which the body is present in the experience of touching my left hand with my right, requires an act of reflection.

[74] Merleau-Ponty notes that in this respect my body is not an object among other objects. It is never in front of me but rather always "with" me. That is, I have a non-thetic awareness of the body at this level.

[75] Husserl (1989, pp. 165–166; 1982, pp. 116–117) also notes that the body is the center of orientation around which the rest of the spatial world groups itself. Consequently, my body has the central mode of givenness of "Here" whereas all other physical things are given as being located "There" in relation to my body. Following Husserl, Schutz (1962d, pp. 222–226), in recognizing the body as the basic scheme of orientation, analyzes the strata of reality in the everyday world in terms of the world within actual reach, and the world within potential and restorable reach. It should be noted, however, that whereas Husserl's reduction to the sphere of "ownness" led him to a "mundanizing apperception" of the body as a special kind of object – the sole object that is immediately and spontaneously governable by the will – Sartre and Merleau-Ponty are concerned to explore the existential relation with body which is revealed in the apprehension that my body is indeed both my center of orientation and the locus of my intentions.

[76] Engelhardt (1973, pp. 41–42) makes the point that "my body appears as that from whence and through which I operate upon the world." The body is the unique vehicle of one's agency and one's most primordial means for contact with the world. Projects are thus always envisaged in terms of one's body. See also Husserl (1989, pp. 159–160).

[77] For example, as Merleau-Ponty (1962, pp. 76 ff.) notes, such phenomena as phantom limb cannot adequately be accounted for in terms of either a purely physiological or a purely psychological explanation. However, they may be understood in terms of the perspective of bodily being-in-the-world. To have a phantom limb is to retain the practical action-oriented field which one enjoyed before mutilation. It is to continue to be "intervolved in a definite environment, to identify oneself with certain projects and be continually committed to them." It is not that the person who has lost a limb merely remembers it, or in some way experiences some sort of "representation" of the absent limb, it is rather that his body remains open to the types of actions for which this limb would be the center if it were still operative (Zaner, 1964, p. 157).

[78] In the case of the phantom limb, the manipulatory movements of a certain part of the body have been destroyed yet the habitual intentions of the lived body remain operative. For a time objects still present themselves to the patient as utilizable and, consequently, he or she constitutes a "fictive body" by means of which he/she continues to aim at the world (Zaner, 1964, p. 157). In this connection it is interesting to note that the phantom limb phenomenon never occurs when amputation is performed just after birth, and is rarely experienced in children in cases where limb amputation occurred before they developed the use and coordination of the limb (Simmel, 1962).

[79] Sacks (1985a, p. 50) notes that this intuitive sense of body is indispensable for our sense of *ourselves*. In his clinical tale, "The Disembodied Lady," he recounts how the loss of this sense caused a patient to experience herself as disembodied and consequently to experience the loss of the "fundamental, organic mooring of (her) identity," her very sense of self. Sacks suggests that corporeal identity is the basis of self and the loss of this corporeal identity deprived his patient of her existential, epistemic, basis.

[80] For a fascinating account of such understanding through gestural display see Sacks (1985d). In this work Sacks describes how aphasic patients who are unable to understand spoken words, as such, may understand much of what is said to them by interpreting extraverbal clues such as expressions, gestures, posture, tone of voice and so forth.

[81] In this respect J. H. Van den Berg (1955, p.42) notes that all so-called "psychical qualities" are *ways in which the body is lived*. Thus, "the voices of aggressive people are hard, their muscles bunched, their blood pulses more fiercely through the vessels," and so forth (emphasis mine).

[82] It should be emphasized, of course, that the apprehension of body-as-object does not require a concrete experience of being looked-at by another person. For example, one is aware of one's body-as-object in moments of shame or humiliation even when there is no actual onlooker present. What Sartre is concerned to argue is that such instances are parasitic on "the Look." That is, when I experience shame or humiliation, I am construing my body as a "being-for-the-Other." I am, so to speak, viewing myself from the point of view of the Other.

For Sartre the experience of "being-for-the-Other" is an essentially negative one. My body as object-for-the-Other manifests itself as facticity rather than subjectivity, as an ensemble of sense organs – "flesh," as an instrument for the Other's "gaze," as transcendence-transcended. However, I would argue (pace Sartre) that under normal circumstances there are some experiences in which the body is objectified as a "being-for-the-Other" which are not perceived as negative in nature. For example, in experiencing myself as a "being-for-the-Other" in the gaze of my lover, I perceive my body-as-object but this may be a positive apprehension of the object-body.

[83] It is obvious that here I part company with Sartre. However, while I do not believe that the objectification of the body arises *solely* in the experience of "being-for-the-Other," I do agree with Sartre that this experience is one of the important ways in which my body is revealed to me as a physical material entity.

[84] This latter experience is particularly alienating. Herbert Plügge (1970, p. 296) notes, for example, that a dead limb "takes on many characteristics of objective thinglikeness, such as an importunate heaviness, burden, weight, with the quality of a substance that feels essentially strange, wooden, like plaster of paris, in any event as largely space-filling and hence not altogether as a part of ourselves." Obviously this sense of alienation is particularly pronounced in pathological disturbances such as paresis. In another context (Toombs, 1991) I have argued that the loss of tactual and kinesthetic sensation is experienced as a radical disengagement of body from self. As Husserl (1989, pp. 152 ff.) points out, kinesthetic sensations not only give the body an "interior" clearly identifying it as mine, but combine with movements in such a way that I experience such movements as my own. To lose this sense is to become disassociated from one's body. It is "the" arm, rather than "my" arm which moves. For a discussion of Husserl's analysis of the role of kinesthetic sensation in the constitution of lived-body see Engelhardt (1977, pp. 51–70).

[85] Obviously this may vary according to culture, time period, and so forth. In China, for example, incorporated into the object-body will presumably be some concept of "qi" (vital energy) and of "yin" and "yang" elements (Kleinman, 1988, p. 109).

[86] Sartre (1956a, pp. 431–432) also refers to the contingent necessity of the body. The body is at once "the necessary condition for the existence of a world and ... the contingent realization of this condition." "We must recognize," he says, "both that it is altogether contingent and absurd that I am a cripple, the son of a civil servant or laborer, irritable and lazy, and that it is nevertheless *necessary* that I be *that* or else something else ..." Insofar as, for Sartre, to be is to choose oneself then I choose the way in which I constitute my contingency (i.e., I may choose to constitute my disability as "unbearable," "the justification for my failures," "fortunate," and so forth). The body (with its contingent necessity) is precisely the necessity that there be a choice at all – "my finitude, my embodiment, is the condition for my freedom."

[87] For a further discussion of this point see Gallagher (1986, pp. 144–146).

[88] This is the case both in acute and chronic illness. In chronic illness the disruption of world is obviously of longer lasting duration and may appear more evident, but even such bodily disorders as a simple cold produce a concurrent disruption of the patient's being-in-the-world. Indeed, the severity of the illness is related to, and judged by, the extent to which the patient's world is disrupted.

[89] As will be noted in the discussion on body-as-object in illness, this forced attention to body engendered at the pre-reflective level results in a shift in attention whereby the body is explicitly thematized and constituted as an object.

[90] Van den Berg (1955, pp. 28–37) gives an excellent discussion of the change in character of the surrounding world which occurs with illness.

[91] That such body style represents one's corporeal identity can be well illustrated by the following quote (cited in Asch and Fine, 1988) from a disabled woman who was born with a disordered corporeal style and who resisted her mother's attempts to make her appear more typical:

She (her mother) made numerous attempts over the years of my childhood to have me go for physical therapy and to practice walking more 'normally' at home. I vehemently refused all her efforts. She could not understand why I would not walk straight.... My disability, with my different walk and talk and my involuntary movements, having been with me all my life was part of me, part of my identity. With these disability features, I felt complete and whole. My mother's attempt to change my walk, strange as it may seem, felt like an assault on myself, an incomplete acceptance of all of me, an attempt to make me over (Rousso, 1984, p. 9).

Since her disordered manner of being did not represent a *change* in body style, she did not perceive it as alien.

[92] For a marvelous account of how cultural attitudes shape the experience of "being different" see Erving Goffman's (1963) *Stigma: Notes on the Management of Spoiled Identity*.

[93] This, of course, relates to Merleau-Ponty's notions regarding the "ambiguity" in the structure of the lived body.

[94] Studies have shown that the views of the non-disabled toward persons with disabilities are overwhelmingly negative (Siller, Ferguson, *et al.*, 1976, cited in Asch and Fine, 1988). Such cultural attitudes necessarily contribute to the loss of self esteem which is an intrinsic element of disability.

[95] This experience invariably occurs when I am traveling through airports. My husband wheels me up to the security barrier, the attendant at the barrier looks directly at me, then turns to my husband and says "Can *she* walk at all?" We now have a standard reply. My husband says, "Yes, and she can talk too!"

[96] This is obviously also the case in the examination room of the doctor's office.

[97] For an excellent discussion on the diminishment of self and the change in social relations which is engendered by loss of upright posture see Murphy (1987).

[98] I am personally aware of the profound effect of loss of upright posture on social interaction whenever I have attended stand-up gatherings (receptions and so forth). In my scooter or wheelchair I am approximately three and a half feet tall and, for the most part, the conversation takes place above my head. When speaking to a standing person, I have to "look up" at them and they "down" at me. This gives me the ridiculous sense of being a child again surrounded by very tall adults. In constantly "looking up" to others and being "looked down on," one feels oneself concretely diminished in person as well as in body. This sense of diminishment is exacerbated by the attitudes of others. One disabled woman reports that whenever she and her husband are shopping and he is pushing her wheelchair, people come up to them and say to him "You must be a saint" (Asch and Fine 1988) – the implication being that she is a burden and he is either saintly or a loser. While this has not happened to me personally, it IS the case that many times people will remark to me "How *lucky* you are to have your husband!" This comment is not uttered as a reflection regarding my

husband's character but is a not-so-oblique reference to my disability – the perception being that since I am disabled, I am a burden and that my relationship with my husband is one of dependence.

[99] As Merleau-Ponty has noted this "seeing through the body" is not limited to the disabled. Under normal circumstances we encounter a narrow passageway through which we must pass not merely as a space with certain measurable dimensions but as a "restrictive potentiality" for the body requiring a modification of our actions. One knows without thinking about it that one must turn sideways in order to proceed past the obstruction. What is peculiar about this "seeing through the body" in disability is that it renders explicit one's being as a being-in-the-world. A problem with the lived body is a problem with the body/environment.

[100] It is worth noting that the restrictive character of physical space for the motor disabled is exacerbated by the fact that until very recently all of our architecture was constructed for people with working legs. Consequently the disabled individual never knows if it will be possible to get into a building (house, cinema, restaurant, motel, office), ride the subway or catch a plane, cross the street and get on to the sidewalk on the other side of the road, and so forth, unless he or she first makes inquiries. More often than not there are physical barriers (steps, escalators, curbs, narrow doorways, uneven surfaces) preventing access.

[101] With regard to the structure of lived temporality, Schutz (1962 f, pp. 306–310) argues that the world is organized around the individual in terms of not only the world within actual reach (including the manipulatory sphere) but also in terms of the world within potential and restorable reach. The world within potential and restorable reach includes past experiences which can be brought back within actual reach through memory, and anticipations for the future based on the individual's experiences of the past and present. As has been noted, the works of Merleau-Ponty, Sartre and Zaner provide important insights into the disintegration of lived spatiality (the world within actual reach) which occurs in illness. Schutz's work suggests that illness may additionally be considered in terms of the disturbance of lived temporality (i.e., with reference to the world within restorable and potential reach). For example, as Sacks (1985b) has shown, the loss of world within restorable reach (through the loss of memory) has profound effects with regard to the loss of self.

[102] A dramatic example of the changed experience of time that can occur with a particular type of illness is one cited by Sacks (1983, pp. 305–306) with regard to a patient with severe Parkinsonism. Sacks relates that the patient appeared to sit absolutely immobile in a wheelchair for fifteen hours at a stretch. The only discernible change was a shift in the location of one hand; in the morning it might be close to one knee, in the middle of the day it might be "frozen" halfway to his nose, and still later the hand might be "frozen" on the patient's nose or glasses. Sacks assumed these were simply "akinetic poses." However, when asked about these movements after successful treatment with L-DOPA, the patient's response was, "What do you mean 'frozen poses'? I was merely wiping my nose." As far as the patient was concerned, these movements had only occupied a single second; he was astonished to discover that they had in fact taken many hours.

[103] It is also the case that illness may engender a *loss* of the past in that degenerative diseases or surgical procedures may cause one to feel that one is losing the body's

continuity with its youth (or former state of being). For instance, Frank notes that he felt such a loss most keenly shortly before surgery to remove a tumorous testicle.

That night I knew that after surgery I would never be the same.... Surgery and chemotherapy would irrevocably break my body's continuity with its past. I did not dread what I would become, but I needed to mourn the end of what I had been.... When you say goodbye to your body ... you say goodbye to how you have lived (Frank, 1991, p. 37).

I would argue that, in this "loss of the past," degenerative diseases and radical surgical procedures represent not simply an episode in the life narrative of the patient but a radical break in that narrative in that they engender a permanent transformation in the patient's being-in-the-world.

[104] For an interesting discussion on the ontology of the body see Pellegrino and Thomasma (1981, pp. 107–108).

[105] There seems to be an important analogy here with Martin Heidegger's (1962, pp. 42–43) analysis of the breakdown of the tool. The breakdown of the tool discloses certain fundamental intentional structures which are normally not rendered explicit. Furthermore, in breakdown the tool becomes conspicuous and obstructive (as does the body in illness).

[106] See the discussion of this point under the constitution of "disease" in Chapter Two.

[107] It should be noted how different this is from the pre-reflective level of lived body wherein the body does not exist *partes extra partes* but rather as an intentional unity. That is, at the level of lived body the diversity of body parts and senses form a systematic unity in the worldly engagement of a subject. For a further elaboration of this point see Rawlinson (1986, pp. 42–43).

[108] Husserl (1989, p. 167) notes that the body is unique in that, whereas I have the freedom to change my position with regard to all other objects, I do not have the power to withdraw myself from my body, nor it from me.

[109] In this regard Engelhardt (1973, p. 41) argues that the organs of the body are differentiated in respect of the sense of being me. Whereas it is possible to recognize a distance between myself and all replaceable organs, the sense of the nervous system is unique in that it cannot be replaced and leave me intact. Indeed, one experiences a radical dependence upon the nervous system. In my essay, "The Body in Multiple Sclerosis," (Toombs, 1992) I have argued that this sense of radical dependence is particularly felt in disorders of the central nervous system in that the patient concretely experiences the inability to disassociate himself/herself from his/her malfunctioning body.

[110] In this respect Sartre's account of the contingent necessity of embodiment may provide an affirmative response for the patient who is faced with chronic, incurable illness. It is vitally important for such patients to feel that there are some aspects of their lives over which they have some control. Sartre's account emphasizes that, although patients may have little or no control over the actual course of the disease process, they always retain the freedom to choose how to respond to their predicament, how to constitute the contingency of their illness. In this regard, Sartre sees

embodiment not only as radical limitation but also as possibility. Furthermore, since *all* embodiment involves radical limitation, limitation due to illness should not be regarded as "fatal" to the ultimate integrity of self. No matter what the extent of the physical limitation, a newly defined self may be constituted. In other words the radical limitation brought about by illness is not unique. The body-in-health is as surely subject to limitation as is the body-in-illness, the contingent necessity of embodiment being what it is.

[111] In discussing one's condition with the physician, the "hidden" presence of the body may be directly explored and discussed. Nevertheless (as was noted in Chapter Two) it remains inapprehensible in fundamental ways. One can, for example, learn that one has a specific disorder in one's bodily organism – a lesion in the carotid artery, for example. Yet, one does not experience the lesion directly. Even when it is visualized by means of an arteriogram and pointed out to the patient, the lesion remains ineffable. The patient experiences only its effects. The carotid artery (and the lesion) represent a hidden and threatening presence.

[112] Sacks (1984) has provided a vivid account of the manner in which body may be experienced as alien and reduced to "objecthood" in his autobiographical account of a leg injury. Interestingly enough I have noticed a similar disassociation from my body as a result of increasing loss of motor control. It is as if my inability voluntarily to control the movement of my legs causes me to feel detached from them and they from me. I notice, for example, that when I sit on a chair and try to raise my legs, I note to myself that *"these"* legs, rather than *"my"* legs, will not move. Since they are no longer under my control, I feel alienated from them.

[113] Obviously this alienation from body resolves itself more readily in acute illness when a return to health has been effected (although it may recur with renewed illness). However, in chronic illness the alienation from body is much more profound since there is no possibility of a return to normal functioning.

[114] Frank (1991) describes this experience when he reports a conversation with his physician following a heart attack.

We talked about my heart as if we were consulting about some computer that was producing errors in the output. "It" had a problem. Our talk was classier than most of the conversations I have with the mechanic who fixes my car, but only because my doctor and I were being vague. He was not as specific as my mechanic usually is. I knew more about hearts than I know about cars, but this engine was inside me, so I was even more reluctant to hear about the scope of the damage (Frank, 1991, p. 10).

[115] For a discussion of the differences between the experience of chronic and acute illness see Jennings, *et al* (1988). The authors note that chronic illness cannot be conceptualized as an aberrant situation that marks a temporary, reversible departure from the person's "normal" state. Consequently, chronically ill patients cannot regard their illness as an "alien presence" within the person but have no choice but to try to integrate their illness into their daily lives. One of the difficulties then is to recognize that one's illness is not an "invader to be defeated" but something to negotiate and live with for the rest of one's life. The authors show that this difference between chronic and acute illness has important implications for the care of the chronically ill.

[116] See previous section; also Sartre (1956a, p. 455).

[117] For an elaboration of this point see Herzlich and Pierret (1987, pp. 69–97).

[118] For a discussion of relevance and the acquisition of knowledge see Schutz and Luckmann (1973, pp. 243–331). See also Chapter One.

[119] Plügge notes that this experience is always felt with pathological disturbances of the body:

If my heart weighs like a stone in my breast, I simultaneously reflectively experience all the more that this heaviness is *my* heart, despite the fact that it acts as if it were something independent and autonomous. Here too a having enters with this into the previously unnoticed being. But over and beyond that the experience also of being had. In pathology anyway it is a rule that everything which I have also has me.

One need but consider the unexpected discovery of an eruption on one's own body. The very instant I notice this rash on me, it has me in its clutches. The dialectics of having and being had is almost always linked with uncertainty, anxiousness, perplexity and incipient reflection, and an effort to clarify the meaning of this exanthema (Plügge, 1970, pp. 305–306).

[120] For a discussion on the manner in which the prevailing biomedical model encourages the view that patients *have* diseases see Engel (1976, pp. 127–131).

[121] For a concrete description of the lived spatiality of the blind person see Hull (1990).

[122] Charles Silberman (1991, p. 14) notes that a recent study found that physicians from teaching hospitals of UCLA and Harvard underestimated or failed to recognize 66 percent of their patients' functional disabilities.

[123] It would appear, however, that this will vary with different cultures although Engel notes that, as far as we have been able to ascertain, the helplessness gesture is universal.

[124] This is not to suggest that people with disabilities do not experience shame when they are alone. Goffman illustrates this well in the following:

When I got up at last ... and had learned to walk again, one day I took a hand glass and went to a long mirror to look at myself, and I went alone. I didn't want anyone to know how I felt when I saw myself for the first time. But there was no noise, no outcry; I didn't scream with rage when I saw myself. I just felt numb. That person in the mirror *couldn't* be me. I felt inside like a healthy, ordinary, lucky person – oh, not like the one in the mirror! Yet, when I turned my face to the mirror there were my own eyes looking back, hot with shame ... when I did not cry or make any sound, it became impossible that I should speak of it to anyone, and the confusion and the panic of my discovery were locked inside me then and there, to be faced alone, for a very long time to come (Goffman, 1963, pp. 7–8).

[125] It is interesting to note that studies indicate that patients suffering from acute diseases are more likely to relinquish control of their bodies to the physician than are those suffering from chronic illnesses. Lidz, et al (1983, pp. 539–543) found that passivity and distance from treatment decisions was typical of acute patients whereas

patients with chronic diseases were much more actively involved in treatment decisions.

[126] Although I do not in this context intend to provide a definition of personhood, it seems clear that our conception of who we are incorporates more than simply the body and includes such facets as the various roles we occupy, our relations with others, our intentional activities, our projects, goals, aspirations, and so forth. In other words, it incorporates our "being-in-the-world."

[127] Essays by physicians who have themselves been ill show that the experience does have an impact on their ability to empathize with patients. The following quote from a clinician writing of his experience of trauma is illustrative of the insight gained.

Throughout the years in medicine, from internship onward, I had always considered myself a sensitive physician. I could communicate with my patients, sense their needs, and attempt to meet them. My experience taught me how less than perfect those perceptions were. I suspect that in order to be an ideally sensitive physician, a treater of patients, one must experience at least some form of illness and undergo some degree of medical care. It is only then that our antennae can pick up and respond to the multitude of signals patients are sending us on a daily basis. These signals center not so much upon our well-tuned abilities as diagnosticians or therapists, but rather upon our abilities as uniquely sensitive and caring human beings.... Being a patient ... allows one to internalize, to make formal and concrete an awareness and response that makes the physician far more capable of dealing with the patient before him (Zaret, 1987, pp. 410–411).

While there is no doubt that undergoing illness in one's own life provides the greatest insight into what it means to be ill, it is possible to develop an empathic understanding through focusing on those experiences in everyday life which provide clues as to the disruption which occurs with bodily (or mental) malfunction and which provide a common ground for developing a shared world of meaning.

[128] A portion of the material included in this section was published in Toombs, "The Meaning of Illness: A Phenomenological Approach to the Patient-Physician Relationship," (1987).

[129] We have noted that the awareness of limited control over the body is evident also in health. However, illness causes one to recognize more forcefully the extent to which one has limited control over the preservation of bodily integrity.

[130] It should be stressed that being anxious, concerned, apprehensive, and so forth, is a response to the loss of certainty and the existential threat posed by illness. Consequently, this is not simply related to the severity of the condition or the critical nature of the therapeutic intervention. As a professor of medicine once remarked to his students, "the only 'minor' surgery is the surgery done on someone else."

[131] Cassell (1985a, p. 30) notes, for example, that the first words of one of his patients found to have carcinoma of the breast were, "I knew it, I'm being punished." Cassell further notes that it is useless simply to tell this patient that carcinoma of the breast is not punishment for some behavior since this aspect of belief and meaning is part of the person. Larry and Sandra Churchill (1989, p. 1127) have pointed out that the notion that illness is related to moral transgression is not uncommon. Indeed, persons

who suffer from cancer of "highly valued portions of the body such as the face or genitals seem especially prone to etiologies in which illness symbolizes retribution for such lapses as excessive vanity or marital infidelity." The authors also argue that to insist that such notions are irrational is irrelevant because patient models cluster around the question of the personal meaning of the disease. Consequently, such notions are not irrational but nonrational. In the case of AIDS, the notion that illness is a direct result of divine punishment is, in fact, quite widespread in our culture (especially perhaps among those who do *not* suffer from the disease).

[132] Indeed, Leder (1990b) suggests such loss of control is global – the individual feels out of harmony with the universe in that he or she has violated the order of the universe.

[133] Sacks (1983) makes the point that this longing for a complete restoration of health is not simply the result of advances in medical science (although I believe such advances raise the patient's expectations that such a restoration is possible).

All of us have a basic, intuitive feeling that once we *were* whole and well; at ease, at peace, at home in the world; totally united with the grounds of our being; and that then we lost this primal, happy, innocent state, and fell into our present sickness and suffering....

We may expect to find such ideas most intense in those who are enduring extremities of suffering, sickness, and anguish, in those who are consumed by the sense of what they have lost or wasted, and by the urgency of recouping before it is too late. Such people, or patients, come to priests or physicians in desperations of yearning, prepared to believe anything for a reprieve, a rescue, a regeneration, a redemption....

This sense of what is lost, and what must be found, is essentially a metaphysical one. If we arrest the patient in his metaphysical search, and ask him *what it is* that he wishes or seeks, he will not give us a tabulated list of items, but will say, simply, 'My happiness,' 'My lost health,' 'My former condition,' 'A sense of reality,' 'Feeling fully alive,' etc. He does not long for this thing or that; he longs for a *general* change in the complexion of things for everything to be *all right* once again, unblemished, the way it once was (Sacks, 1983, p. 27).

[134] For that matter the transformation to "objecthood" as a direct result of illness may also manifest itself in the "gaze" of strangers, or even family members, to the extent that the overt signs of illness draw explicit attention to the malfunctioning object body.

[135] Pellegrino (1979b, p. 45) argues that the vulnerability engendered by physical illness is unique in that our capacity to deal with it is severely impaired. That is, whereas other conditions (such as imprisonment, economic deprivation, and so forth) deprive one of the freedom to act, we usually feel that we can cope with these other states of vulnerability if only we have our "health." Health is perceived as a means towards freedom and other primary values.

[136] For a revealing autobiographical account illustrating this point see Rosenbaum (1988). Dr. Rosenbaum also notes that, as a patient, he acquiesced to his therapist even though what she told him contradicted his own experience.

I really didn't notice any difference in the muscles, but I didn't want to hurt her feelings. That was dumb, but it is a common reaction of patients: don't upset the doctor or he or she won't like you and won't take good care of you. I was startled by that insight. Had I known that before? (Rosenbaum, 1988, p. 26).

[137] For an excellent discussion on the whole question of uncertainty in medicine see, Silberman, *Crisis in Medicine* (forthcoming); also Katz (1984, pp. 165–206).

[138] In this regard Pellegrino (1979a, pp. 169–194) argues that a clinical judgment is made up of three generic questions: What can be wrong? What can be done? What should be done? He argues that "a right healing action for a particular patient" frequently involves the counterposition of what is good scientifically, what the physician thinks is good, and what the patient will accept as good. The answer to the question, "What should be done?" depends on a myriad of factors in the patient's life situation and, particularly, must take into account the patient's value system.

[139] Sartre's (1956b) short story, "The Wall," well illustrates this point.

[140] It is important to note that these characteristics are not simply "psychological effects" accompanying a disease state – and thus the province of the psychologist rather than the physician. For instance, "loss of the familiar world" (and the disruption of space and time) is not a "psychological" problem. Rather, these losses are integral elements of the human experience of illness and they must be attended to as a part of that experience.

[141] For discussions on the nature of empathy see Husserl (1989, pp. 170–180; pp. 208 ff.; 239 ff.; 362 ff.; 381 ff.); Scheler (1970); Stein (1970); Zaner (1981, pp. 181–241). For a discussion of the nature of empathy and its use in clinical situations see Katz (1963).

[142] It should be noted that under these circumstances, although the body is understood as a purely physiological entity, it is less likely to be experienced as "uncanny." Rather, in focusing on the mechanistic processes of his or her physical body, the medical student apprehends it as simply an exemplar of THE human body without necessarily thereby experiencing the existential alienation from self. Of course, existential anxiety can arise from this recognition of the body as a purely physiological entity. Hence, the susceptibility of some medical students to suspect that they are suffering from one or other of the various disorders that they are studying ("medical students' neurosis"). In his book, *A Coronary Event*, Michael Halberstam (a practicing internist and cardiologist) makes the interesting observation that whereas such existential anxiety on the part of medical students is relatively rare, it is common among practicing physicians:

It takes only a few years of practice for physicians to become uneasily aware that death and illness are not things that happen to others but the condition of all mankind.... After all, thirty-five and forty year old lawyers and professors and farmers die suddenly all the time, and every practicing physician becomes painfully aware of this. Week after week he listens to patients his own age tell how they noticed the first slight chest tightness that finally ended in crushing chest pain. Day after day he visits heart patients his own age in the hospital, watching their pulse rate and blood pressure being kept normal by the grace of medication alone. Hour after hour the doctor reads the medical literature about stress, about hard work, about cholesterol, about the coronary artery disease that was found already developing among eighteen– and twenty-year old soldiers killed in Korea. Minute after minute he feels the beat of his own pulse, the throb of his own blood pressure, the silent tides of life. Like a mechanic whose ears can pick up the street sound of a faulty carburetor in the middle of a Sunday sermon, the physician is always half tuned to the workings of his own body, half ready for the pain and pressure in his own chest that he has heard described a hundred times over by his patients.

... A pretty unmarried nurse I know works in the coronary care unit of a big hospital. For a while she went out with a succession of cardiologists and cardiology residents until she found the whole routine a bit frightening. They all had, she explained, a tendency during moments of passion to experience palpitations, skipped beats, rapid pulse and headache, and to translate these evidences of desire into signs of a coronary (Halberstam and Lesher, 1976, pp. 42–43).

This book gives a fascinating narrative account of a heart attack from the point of view of the patient who suffered the attack (Stephan Lesher) and the doctor who treated him (Michael Halberstam).

[143] In explicitly attending to the lived experience of the patient in terms of his or her pre-scientific understanding of everyday life, the physician is doing so within the natural attitude prior to interpreting that experience in terms of the naturalistic attitude.

[144] This is in no way to suggest that it is possible fully to grasp the meaning of another's experience, nor is it to minimize the extent to which the lived experience of the body changes in illness. Sacks (1984, p. 202) notes, for example, that though he tried with all the imagination and empathy he could muster to enter the experience of his patients with Parkinsonism he was finally unable to do so. One cannot, he says, "imagine" Parkinsonism without *being* Parkinsonian. My suggestion is however, that whereas one cannot fully grasp the experience of being Parkinsonian, one can "imagine" something (albeit incomplete) of the profound threat to one's being that such a disorder represents. And one can do this specifically because one is provided with clues in the lived experience of the body.

[145] Indeed, Silberman (1991, p. 15) quotes a study by Beckman and Frankel that found that, on average, physicians interrupted patients 18 seconds after the patient began to speak and patients were able to complete their statements in only 23 percent of the visits.

[146] In this regard, Cassell (1985a, p. 15) notes that hospitalized patients (particularly in a teaching hospital) have been "trained" to tell their story (or give the history) in the way physicians prefer to hear it. That is, patients tend to emphasize what doctors seem to find important rather than what is of personal significance.

[147] In this respect Husserl has noted that one may distinguish between the thematic attitude directed at the "objective" world as scientific theme (the "scientific attitude") and the "personal attitude." In the "personal attitude" attention is directed to the meaning that an individual's experience has for him personally:

[I[n the personal attitude, interest is directed toward the persons and their comportment toward the world, toward the ways in which the thematic persons have consciousness of whatever they are conscious of as existing for them, and also toward the particular objective sense the latter has in their consciousness of it. In this sense what is in question is not the world as it actually is but the particular world which is valid for the persons, the world appearing to them with the particular properties it has in appearing to them; the question is how they, as persons, comport themselves in action and passion – how they are motivated to their specifically personal acts of perception, of remembering, of thinking, of valuing, of making plans, of being frightened and automatically starting, of defending themselves, of attacking, etc. Persons are motivated only by what they are conscious of and in virtue of the way in which this (object of consciousness) exists for them in their consciousness of it, in virtue of its sense – how it is valid or not valid for them, etc. (Husserl, 1970a, p. 317).

In explicitly attending to the patient's story of illness, the physician is adopting the "personal attitude."
[148] As was noted in Chapter Two, at the level of "disease" the patient assigns explanatory meaning to the lived experience of illness. Nevertheless, it will be recalled that the patient's conception of the illness in terms of "disease" is significantly different from the physician's concept of the patient's illness as a "disease state." Consequently, the explanatory model of the patient is not identical with the physician's explanatory model.
[149] In this respect it is interesting to take note of an educational endeavor which has been instituted by Dr. Rita Charon at Columbia University. Charon directs a course for second year medical students which explicitly attempts to teach the "empathic stance." The course introduces students to the voice and the world of the patient – the challenge being to "coax medical students away from their detached objectifying stance, a stance inevitably produced in them through the reductionism of most of their curriculum, without disarming them or rendering them ineffective through over-identification with the patient" (Charon, 1989, p. 139). An important exercise in the course is to have students interview a patient with chronic illness. The patient is invited to tell the story of his or her illness and the students are directed to focus on the patient's own understanding of the illness and to learn how it has changed the patient's life. After the interview students are required to write an account of the patient's illness using the narrative voice of the patient. In asking them to recognize and adopt the patient's voice through writing, Charon notes that she is asking them to seek out the patient's perspective, the coherence that a patient visits on a set of events, and ultimately the meaning that the patient attaches to it all. This exercise accomplishes many goals: students comment that writing the stories allows them to feel something of the patient's experience, and comparing the stories written from the same interview demonstrates to them the selectiveness of attention and the personal contribution of the hearer. Most importantly, students become more attuned to the patient's apprehension of illness, to the life context within which the illness takes place, to the ordeal of being sick, and to the many ways there are to heal. As Charon puts it:

The imagination is a powerful instrument in the practice of medicine. The physician's effectiveness increases with empathy, and empathy springs from the ability to imagine the patient's point of view. This encounter hinges on narrative acts; on the patient's ability to tell a story, and on the interviewer's skill in receiving it and hearing its message (Charon, 1989, p. 137).

Charon's point is that the physician's effectiveness qua physician increases with empathic understanding. It is not simply that such understanding insures collegiality between doctor and patient (i.e., the patient is treated in his uniqueness as an individual) but that understanding the patient's illness in the context of his particular life situation enables the physician to devise maximally effective therapeutic measures.

[150] For an illustration of this contrast see, James H. Buchanan, *Patient Encounters: The Experience of Disease*, 1989. In this book Buchanan focuses on 16 diseases not only as they are diagnosed and treated, but as they are suffered by individuals. Prior to telling each illness story, Buchanan sets forth the textbook description of the disease state. The patients' narratives clearly show that the biomedical description captures nothing of the illness experience or the suffering which accompanies it.

[151] Schutz (1962f, p. 318) notes that there are, of course, other social relations (such as the world of my contemporaries, the world of my predecessors and the world of my successors) but he argues that the face-to-face relation is the most central dimension of the social world.

[152] Schutz notes that the Other is from the outset given to me as both a material object with its position in space and as a subject with its psychological life (the psychological life being apprehended rather than given in an originary presence). For a discussion on the appresentation of the Other see Husserl (1989, pp. 170 ff.).

[153] It should, of course, be noted that in the case of preventive medicine (where patients do not come to the physician because they are experiencing illness) the relationship takes on a different character. In this case the end of the patient-physician relationship is not a restoration of bodily integrity but the maintenance of health. I would argue, however, that this type of physician-patient relationship is a derivative relationship and the primary physician-patient relationship is the one so described.

[154] This difference in emphasis is well reflected in the following remarks by the physician, David Rabin, writing of his experience of amyotrophic lateral sclerosis:

I traveled to a prestigious medical center renowned for its experience with ALS. The diagnostic and technical skills of the people were superb, and more than matched the reputation of the institution. The neurologist was rigorous in his examination and deft in reaching an unequivocal diagnosis. My disappointment stemmed from his impersonal manner. He exhibited no interest in me as a person, and did not make even a perfunctory inquiry about my work. He gave me no guidelines about what I should do, either concretely – in terms of daily activity – or, what was more important, psychologically, to muster the emotional strength to cope with a progressive degenerative disease.... The only thing my doctor did offer me was a pamphlet setting out in grim detail the future that I already knew about too well.... I still recall that the only time he seemed to come alive during our interview was when he drew the mortality curve among his collected patients for me. "Very interesting," he said, "there's a break in the slope after three years." (Rabin, 1985a, p. 32).

155 As Martha Weinman Lear notes of her physician-husband fighting his own heart disease:

[I]t was the medical voice coming through. It was the neat surgical mind demanding an adversary, an enemy, a pathology, recognizable forces of death and disease against which he might pit his own skills.

They were trained like that, to anthropomorphize disease. Some diseases were enemies you could not vanquish: terminal cancers, inexorable progressions downward. Others were mischievous little bastards – sleepers, simple prostates and kidney stones that should have been an easy win but might put up a hell of a fight, even to the death.

"[W]here is my adversary?" It was not a *thing*, not a germ, not a kidney stone, not a cancer or an infection. It was simply this process which was wearing him out, filling him with pain and frustration, and he wanted to fight it aggressively, as he had been trained to do (Lear, 1980, p. 152).

156 Indeed, McWhinney (1978, p. 299) argues that the emphasis on "objective" knowledge in medicine and the dominance of mechanistic values over other values in medicine results in such consequences as inappropriate routines of investigation, unnecessary precision, spurious objectivity, redundant investigation, selective inattention to information and inappropriate standardization. He suggests that, whereas there are undoubtedly benefits to advanced technology, technology is harmful when its values override other human values without any substantial net benefit.

157 In writing of her experience in medical school, Perri Klass makes the interesting observation that clinical vernacular reveals the tendency to blame the patient (and not the limits of medical science) for any failure to respond to treatment.

You never say that a patient's blood pressure fell or that his cardiac enzymes rose. Instead a patient is always the subject of the verb: "He dropped his pressure." "He raised his enzymes." ... When chemotherapy fails to cure Mrs. Bacon's cancer, what we say is, "Mrs. Bacon failed chemotherapy." (Klass, 1987, p. 72).

For further elaboration of this point see Donnelly (1986, pp. 81–94).

158 As a person living with incurable illness, and more particularly with multiple sclerosis, I can attest to the fact that the physician's participation is crucial in assisting the patient to retain control and cope with the realities of his illness. Without such participation on the part of the physician, the patient often feels helpless in the face of circumstance.

159 Jean Craig's (1991) account of her husband's fatal illness vividly shows how vital it is that the physician be supportive of the dying patient (and how devastating it is when the patient senses the physician's loss of interest).

What I'm after is the right to die with people who are on your team....

What I was fighting for – asking for – was human dignity. What I wanted for Ed was so simple. A doctor who would help him die well. Who would offer him not hope, but trust and confidence. A doctor who would say, "I'm still with you, Ed." (Craig, 1991, p. 309).

It is extremely important for chronically ill and dying patients to feel they are not "medically" alone. Even if the physician cannot "do anything" in the sense of intervening in such a way as to eradicate or halt the disease process, he or she can support the patient's struggle. Too often "hope" is equated with the goal of "cure" and thus patients who cannot be "cured" are deemed "hopeless."

[160] In contrast, says Toulmin (1976, pp. 46–47), as biomedical scientist the physician's understanding, like all scientific understanding, will remain entirely general: "His questions – qua scientific – are entirely general questions about THE brain, THE liver, etc..... This being so, his interest in particular patients will be minimal and accidental: the more of his research he can do with laboratory animals or *in vitro*, the better." Engelhardt (1982, p. 53) argues that we must make a distinction between the basic sciences as very successful explanatory and predictive exercises in their own right and basic sciences as auxiliary to the social goals and individual interests that direct medicine as an applied science. He suggests that both the clinical approach and the basic scientific approach are onesided and incomplete – that is, each requires the other. Medicine developed explanatory models in order better to treat patients' complaints – the former being secondary to the latter.

[161] Pellegrino (1979a), Cassell (1977, 1982) and others (Engel, 1987; McWhinney, 1986) argue that indeed medicine has two obligations: the relief of human suffering and the prolongation of human life. Since the latter is not always possible, the former is equally important.

# BIBLIOGRAPHY

Asch, A. and Fine, M.: 1988, 'Introduction: Beyond pedestals', in A. Asch and M. Fine (eds.), *Women and Disabilities*, Temple University Press, Philadelphia, pp. 1–37.

Alsop, S.: 1973, *Stay of Execution*, J.B. Lippincott Company, Philadelphia, Pennsylvania.

Bachelard, S.: 1968, *A Study of Husserl's Formal and Transcendental Logic*, Lester Embree (trans.), Northwestern University Press, Evanston, Illinois.

Baron, R.J.: 1985, 'An introduction to medical phenomenology: I can't hear you while I'm listening', *Annals of Internal Medicine* **103**, 606–611.

Baron, R.J.: 1981, 'Bridging clinical distance: An empathic rediscovery of the known', *The Journal of Medicine and Philosophy* **6**, 5–23.

Bass, M.J. *et al.*: 1983, 'The natural history of headache in family practice', paper presented at the World Organization of National Colleges and Academies of General Practitioners, Singapore.

Bell, T.D.: 1991, 'Search for wholeness: The adventures of a doctor-patient', *The Pharos*, Winter, 29–30.

Bichat, X.: 1981, 'Pathological anatomy, preliminary discourse', in A.L. Caplan *et al.* (eds.), *Concepts of Health and Disease: Interdisciplinary Perspectives*, Addison-Wesley, Reading, Massachusetts, pp. 167–173.

Blacklock, S.: 1977, 'The symptom of chest pain in family practice', *The Journal of Family Practice* **4**, 429–433.

Bosch, G.: 1970, *Infantile Autism: A Clinical and Phenomenological Investigation Taking Language as a Guide*, D. Jordan *et al.* (trans.), Springer Verlag, New York.

Brody, H.: 1987, *Stories of Sickness*, Yale University Press, New Haven, Connecticut.

Bruner, J.: 1987, 'Life as narrative', *Social Research* **54**, 11–32.

Buchanan, J.H.: 1989, *Patient Encounters: The Experience of Disease*, University Press of Virginia, Charlottesville, Virginia.

Carr, D.: 1986, *Time, Narrative and History*, Indiana University Press, Bloomington, Indiana.

Carson, R.: 1986, 'Care and research: Antinomy or complement', in J. Van Eys *et al.* (eds.), *The Common Bond: The UT System Cancer Center Code of Ethics*, Charles C. Thomas, Springfield, Illinois, pp. 47–55.

Casey, E.S.: 1976, *Imagining: A Phenomenological Study*, Indiana University Press, Bloomington, Indiana.

Casey, E.S.: 1977, 'Imagination and phenomenological method,' in F.A. Elliston *et al.* (eds.), *Husserl: Expositions and Appraisals*, University of Notre Dame Press, Notre Dame, Indiana, pp. 70–82.

Cassell, E.J.: 1985a, *Clinical Technique*, Vol. 2 of *Talking With Patients*, The MIT Press, Cambridge, Massachusetts.

Cassell, E.J.: 1976, 'Illness and disease', *Hastings Center Report* **6**, 27–37.
Cassell, E.J.: 1986, 'Quality of life is a personal choice', in J. van Eys *et al.* (eds.), *The Common Bond: The U.T. System Cancer Center Code of Ethics*, Charles C. Thomas, Springfield, Illinois, pp. 57–65.
Cassell, E.J.: 1977, 'The function of medicine', *Hastings Center Report*, 16–19.
Cassell, E.J.: 1966, *The Healer's Art*, J.B. Lippincott, New York.
Cassell, E.J.: 1982, 'The nature of suffering and the goals of medicine', *The New England Journal of Medicine* **306**, 639–645.
Cassell, E.J.: 1991, *The Nature of Suffering*, Oxford University Press, New York.
Cassell, E.J.: 1991, 'Recognizing suffering', *Hastings Center Report* **21**, 24–31.
Cassell, E.J.: 1979, 'The subjective in clinical judgment', in H. T. Engelhardt, Jr. *et al.* (eds.), *Clinical Judgment: A Critical Appraisal*, D. Reidel, Dordrecht, Holland, pp. 199–215.
Cassell, E.J.: 1985b, *The Theory of Doctor-Patient Communication*, Vol. 1, *Talking With Patients*, The MIT Press, Cambridge, Massachusetts.
Cassell, E.J.: 1983, letter to the author.
Charon, R.: 1989, 'Doctor-patient/reader-writer: Learning to find the text', *Soundings* **72**, 137–152.
Churchill, L. and Churchill, S.: 1989, 'Storytelling in medical arenas: The art of self determination', *Journal of the American Medical Association* **262**, 1127.
Cohen, H.: 1981, 'The evolution of the concept of disease', in A. L. Caplan *et al.* (eds.), *Concepts of Health and Disease: Interdisciplinary Perspectives*, Addison-Wesley, Reading, Massachusetts, pp. 209–219.
Craig, J.: 1991, *Between Hello and Goodbye*, Jeremy P. Tarcher, Inc., Los Angeles.
Donnelly, W.: 1986, 'Medical language as symptom: Doctor talk in teaching hospitals', *Perspectives in Biology and Medicine* **30**, 81–94.
Elder, A. and Samuel, O. (eds.): 1987, *'While I'm Here, Doctor': A Study of Change in the Doctor-Patient Relationship*, Tavistock Publications, New York.
Engel, G.L.: 1985, 'Commentary on Schwartz and Wiggins: Science, humanism, and the nature of medical practice', *Perspectives in Biology and Medicine* **28**, 362–365.
Engel, G.L.: 1987, 'Physician-scientists and scientific physicians: Resolving the humanism-science dichotomy', *The American Journal of Medicine* **82**, 107–111.
Engel, G.L.: 1977a, 'The care of the patient: Art or science?', *The Johns Hopkins Medical Journal* **140**, 222–232.
Engel, G.L.: 1977b, 'The need for a new medical model: A challenge for biomedicine', *Science* **196**, 129–136.
Engel, G.L.: 1976, 'Too little science. The paradox of modern medicine's crisis', *The Pharos*, 127–131.
Engelhardt, H.T., Jr.: 1977, 'Husserl and the mind-body relation', in Don Ihde *et al.* (eds.), *Interdisciplinary Phenomenology*, Martinus Nijhoff, The Hague, pp. 51–70.
Engelhardt, H.T., Jr.: 1976, 'Ideology and etiology', *The Journal of Medicine and Philosophy* **1**, 256–268.
Engelhardt, H.T., Jr.: 1982, 'Illnesses, diseases, and sicknesses', in V. Kestenbaum (ed.), *The Humanity of the Ill*, University of Tennessee Press, Knoxville, Tennessee, pp. 142–56.

Engelhardt, H.T., Jr.: 1973, *Mind-Body: A Categorial Relation*, Martinus Nijhoff, The Hague.

Engelhardt, H.T., Jr.: 1989, 'Pain, suffering, addiction and cancer', in C. Stratton Hill, Jr. *et al.* (eds.), *Drug Treatment of Cancer Pain in a Drug Oriented Society*, Vol. 11, *Advances in Pain Research and Therapy*, Raven Press, New York, pp. 71–79.

Engelhardt, H.T., Jr.: 1974, 'The disease of masturbation: Values and the concept of disease', *Bulletin of the History of Medicine* **48**, 234–248.

Engelhardt, H.T., Jr.: 1982, 'The subordination of the clinic', in B. Gruzalski *et al.* (eds.), *Value Conflicts in Health Care Delivery*, Ballinger Publishing, Cambridge, Massachusetts, pp. 41–57.

Feinstein, A.R.: 1967, *Clinical Judgment*, Robert E. Krieger Publishing Company, Huntingdon, New York.

Foucault, M.: 1975, *The Birth of the Clinic: An Archaeology of Medical Perception*, A.M. Sheridan Smith (trans.), Vintage Books, New York.

Frank, A. W.: 1991, *At the Will of the Body*, Houghton Mifflin Company, Boston.

Gallagher, S.: 1986, 'Lived body and environment', *Research in Phenomenology* **16**, 139–170.

Goffman, E.: 1963, *Stigma: Notes on the Management of Spoiled Identity*, Prentice-Hall, Inc., New Jersey.

Halberstam, M.J. and Lesher, S.: 1976, *A Coronary Event*, J. B. Lippincott, Philadelphia.

Hanna, W.J. and Rogovsky, B.: 1986, 'Women and disability: Stigma and "the third factor."' Unpublished paper, Department of Family and Community Development, College of Human Ecology, University of Maryland, College Park.

Heelan, P.: 1973, 'Hermeneutics of experimental science in the context of the life-world', in D. Ihde *et al.* (eds.), *Interdisciplinary Phenomenology*, Martinus Nijhoff, The Hague, pp. 7–50.

Heidegger, M.: 1962, *Being and Time*, J. Macquarrie *et al.* (trans.), Harper and Row, New York.

Herzlich, C. and Pierret, J.: 1987, *Illness and Self in Society*, E. Forster (trans.), The Johns Hopkins University Press, Baltimore, Maryland.

Hull, J.M.: 1990, *Touching the Rock: An Experience of Blindness*, Pantheon Books, New York.

Husserl, E.: 1970a, 'Appendix III: The attitude of natural science and the attitude of humanistic science. Naturalism, dualism and psychophysical psychology', in D. Carr (trans.), *The Crisis of European Sciences and Transcendental Phenomenology: An Introduction to Phenomenological Philosophy*, Northwestern University Press, Evanston, Illinois, pp. 315–383.

Husserl, E.: 1982, *Cartesian Meditations: An Introduction to Phenomenology*, D. Cairns (trans.), 7th Impression, Martinus Nijhoff, The Hague.

Husserl, E.: 1962, *Ideas: General Introduction to Pure Phenomenology*, W.R.B. Gibson (trans.), Collier Books, London.

Husserl, E.: 1989, *Ideas Pertaining to a Pure Phenomenology and to a Phenomenological Philosophy: Studies in the Phenomenology of Constitution*, R. Rojcewicz and A. Schuwer (trans.), Kluwer Academic Publishers, Dordrecht, The Netherlands.

Husserl, E.: 1929, 'Phenomenology', in *Encyclopaedia Britannica*, 14th edition, Encyclopaedia Britannica Inc., New York, pp. 699–702.

Husserl, E.: 1970b, *The Crisis of European Sciences and Transcendental Phenomenology: An Introduction to Phenomenological Philosophy*, D. Carr (trans.), Northwestern University Press, Evanston, Illinois.

Husserl, E.: 1964, *The Phenomenology of Internal Time-Consciousness*, J.S. Churchill (trans.), Indiana University Press, Bloomington, Indiana.

Jennings, B. *et al.*: 1988, 'Ethical challenges of chronic illness', *The Hastings Center Report*, 1–16.

Katz, J.: 1984, *The Silent World of Doctor and Patient*, The Free Press, New York.

Katz, R.L.: 1963, *Empathy: Its Nature and Uses*, The Free Press of Glencoe, London.

Kestenbaum, V.: 1982a, 'The experience of illness', in. V. Kestenbaum (ed.), *The Humanity of the Ill*, University of Tennessee Press, Knoxville, Tennessee, pp. 3–38.

Kestenbaum, V. (ed.): 1982b, *The Humanity of the Ill, Phenomenological Perspectives*, The University of Tennessee Press, Knoxville, Tennessee.

Klass, P.: 1987, *A Not Entirely Benign Procedure*, Signet Books, New York.

Kleinman, A.: 1988, *The Illness Narratives: Suffering, Healing and the Human Condition*, Basic Books, New York.

Kleinman, A. and Mendelsohn, E.: 1979, 'Systems of medical knowledge: A comparative approach', *The Journal of Medicine and Philosophy* 3, 314–330.

Kohak, E.: 1978, *Idea and Experience: Edmund Husserl's Project of Phenomenology in Ideas I*, University of Chicago Press, Chicago, Illinois.

Konner, M.: 1987, *Becoming a Doctor: A Journey of Initiation in Medical School*, Viking Penguin, New York.

Kravetz, R.E.: 1987, 'Bleeding ulcer', in H. Mandell *et al.* (eds.), *When Doctors Get Sick*, Plenum Publishing, New York, pp. 429–437.

Kuhn, T.S.: 1970, *The Structure of Scientific Revolutions*, 2nd Edition, University of Chicago Press, Chicago, Illinois.

Lauritsen, K. *et al.*: 1985, 'Effect of omperazole and cimetidine on duodenal ulcer: A double-blind comparative trial', *New England Journal of Medicine* 312, 958–961.

Lear, M.W.: 1980, *Heartsounds*, Simon and Schuster, New York.

Leder, D.: 1990a, 'Clinical interpretation: The hermeneutics of medicine', *Theoretical Medicine* 11, 9–24.

Leder, D.: 1990b, 'Illness and exile: Sophocles' Philoctetes', *Literature and Medicine* 9, 1–11.

Leder, D.: 1984a, 'Medicine and the paradigms of embodiment', *The Journal of Medicine and Philosophy* 9, 29–43.

Leder, D.: 1984b, 'Toward a phenomenology of pain', *Review of Existential Psychology and Psychiatry* 19, 255–266.

Leigh, H. and Reiser, M.F.: 1980, *The Patient, Biological, Psychological and Social Dimensions of Medical Practice*, Plenum Publishers, New York.

Lidz, C. *et al.*: 1983, 'Barriers to informed consent', *Annals of Internal Medicine* 99, 539–543.

Luria, A.R.: 1972, *The Man With a Shattered World: The History of a Brain Wound*, L. Solotaroff (trans.), Harvard University Press, Cambridge, Massachusetts.

Luria, A.R.: 1987, *The Mind of a Mnemonist: A Little Book About a Vast Memory*, L. Solotaroff (trans.), Harvard University Press, Cambridge, Massachusetts.

MacIntyre, A.: 1981, *After Virtue*, University of Notre Dame Press, Notre Dame, Indiana.

Mandell, H. and Spiro, H. (eds.): 1987, *When Doctors Get Sick*, Plenum Publishing, New York.

McCullough, L.B.: 1981, 'Thought-styles, diagnosis and concepts of disease: Commentary on Ludwik Fleck', *The Journal of Medicine and Philosophy* 6, 257–261.

McWhinney, I.R.: 1986, 'Are we on the brink of a major transformation of clinical method?' *Canadian Medical Association Journal* 135, 873–878.

McWhinney, I.R.: 1983, 'Changing models: The impact of Kuhn's theory on medicine', *Family Practice* 1, 3–8.

McWhinney, I.R.: 1978, 'Medical knowledge and the rise of technology', *The Journal of Medicine and Philosophy* 3, 293–304.

Merleau-Ponty, M.: 1964, 'Phenomenology and the sciences of man', in J.M. Edie (ed.), *The Primacy of Perception*, Northwestern University Press, Evanston, Illinois, pp. 43–95.

Merleau-Ponty, M.: 1962, *Phenomenology of Perception*, C. Smith (trans.), Routledge and Kegan Paul, London.

Mishler, E.G.: 1984, *The Discourse of Medicine: Dialectics of Medical Interviews*, Ablex Publishing, New Jersey.

Morgagni, G. B.: 1981, 'The seats and causes of disease: Author's preface', in A.L. Caplan *et al.* (eds.), *Concepts of Health and Disease: Interdisciplinary Perspectives*, Addison-Wesley, Reading, Massachusetts, pp. 157–165.

Mullan, F.: 1975, *Vital Signs, A Young Doctor's Struggle With Cancer*, Farrar, Straus and Giroux, New York.

Murphy, R.F.: 1987, *The Body Silent*, Henry Holt, New York.

Natanson, M.: 1973, *Edmund Husserl: Philosopher of Infinite Tasks*, Northwestern University Press, Evanston, Illinois.

Natanson, M.: 1962, 'Introduction', in M. Natanson (ed.), *The Problem of Social Reality*, Vol. 1, *Alfred Schutz: Collected Papers*, Martinus Nijhoff, The Hague, pp. xxv-xlvii.

Natanson, M.: 1968, *Literature, Philosophy, and the Social Sciences: Essays in Existentialism and Phenomenology*, Martinus Nijhoff, The Hague.

Natanson, M.: 1969, 'Philosophy and psychiatry', in E.W. Straus *et al.* (eds.), *Psychiatry and Philosophy*, Springer-Verlag, New York, pp. 85–110.

Netsky, M.D.: 1976, 'Dying in a system of 'good care': Case report and analysis', *Pharos*, 57–61.

Nolan, C.: 1987, *Under the Eye of the Clock*, St. Martin's Press, New York.

Percy, W.: 1954, *The Message in the Bottle*, Straus and Giroux, New York.

Pellegrino, E.D. and Thomasma, D.C.: 1981, *A Philosophical Basis of Medical Practice: Toward a Philosophy and Ethic of the Healing Professions*, Oxford University Press, New York.

Pellegrino, E.D.: 1982, 'Being ill and being healed: Some reflections on the grounding of medical morality', in V. Kestenbaum (ed.), *The Humanity of the Ill*, University of Tennessee Press, Knoxville, Tennessee, pp. 157–166.

Pellegrino, E.D.: 1979a, 'The anatomy of clinical judgments: Some notes on right reason and right action', in H.T. Engelhardt, Jr. *et al.* (eds.), *Clinical Judgment: A Critical Appraisal*, D. Reidel Publishing, Dordrecht, The Netherlands, pp. 169–194.

Pellegrino, E.D.: 1983, 'The healing relationship: The architectonics of clinical medicine', in E. Shelp (ed.), *The Clinical Encounter: The Moral Fabric of the Patient-Physician Relationship*, D. Reidel Publishing, Dordrecht, The Netherlands, pp. 153–172.

Pellegrino, E.D.: 1979b, 'Toward a reconstruction of medical morality: The primacy of the act of profession and the fact of illness', *The Journal of Medicine and Philosophy* **4**, 32–56.

Peterson, W.L. *et al.*: 1977, 'Healing of duodenal ulcer with an antacid regimen', *New England Journal of Medicine* **297**, 341–345.

Piaget, J.: 1970, *Genetic Epistemology*, E. Duckworth (trans.), Columbia University Press, New York.

Plügge, H.: 1970, 'Man and his body', in S.F. Spicker (ed.), *The Philosophy of the Body: Rejections of Cartesian Dualism*, Quadrangle Books, Chicago, pp. 293–211.

Rabin, D.: 1982, 'Occasional notes: Compounding the ordeal of ALS: Isolation from my fellow physicians', *The New England Journal of Medicine* **307**, 506–509.

Rabin, D. *et al.*: 1985a, 'Compounding the ordeal of ALS: Isolation from my fellow physicians', in P.L. Rabin *et al.* (eds.), *To Provide Safe Passage: The Humanistic Aspects of Medicine*, Philosophical Library, New York, pp. 29–37.

Rabin, D. and Rabin, P. (eds.): 1985b, *To Provide Safe Passage: The Humanistic Aspects of Medicine*, Philosophical Library, New York.

Rawlinson, M.C.: 1982, 'Medicine's discourse and the practice of medicine', in V. Kestenbaum (ed.), *The Humanity of the Ill*, The University of Tennessee Press, Knoxville, Tennessee, pp. 69–85.

Rawlinson, M.C.: 1986, 'The sense of suffering', *The Journal of Medicine and Philosophy* **11**, 39–62.

Rosenbaum, E.E.: 1988, *A Taste of My Own Medicine: When the Doctor is the Patient*, Random House, New York.

Rousso, H.: 1984, 'Fostering healthy self-esteem', *The Exceptional Parent*, 9–14.

Sacks, O.: 1983, *Awakenings*, E. P. Dutton, New York.

Sacks, O.: 1984, *A Leg to Stand On*, Summit Books, New York.

Sacks, O.: 1986, 'Clinical tales', *Literature and Medicine* **5**, 14–19.

Sacks, O.: 1985a, 'The disembodied lady', in *The Man Who Mistook His Wife for a Hat and Other Clinical Tales*, Summit Books, New York, pp. 42–52.

Sacks, O.: 1985b, 'The lost mariner', in *The Man Who Mistook His Wife for a Hat and Other Clinical Tales*, Summit Books, New York, pp. 22–41.

Sacks, O.: 1985c, *The Man Who Mistook His Wife for a Hat and Other Clinical Tales*, Summit Books, New York.

Sacks, O.: 1985d, 'The president's speech', *The New York Review of Books*, 29.

Sartre, J.P.: 1956a, *Being and Nothingness: A Phenomenological Essay on Ontology*, H.E. Barnes (trans.), Pocket Books, New York.

Sartre, J.P.: 1956b, 'The wall', in W. Kaufmann (ed.) *Existentialism from Dostoevsky to Sartre*, Meridian Books, Cleveland, Ohio, pp. 223–240.

Scarry, E.: 1985, *The Body in Pain*, Oxford University Press, New York.

Scheler, M.: 1970, *On the Nature of Sympathy*, P. Heath (trans.), Archon Books, Hamden, Connecticut.

Schrag, C.: 1982, 'Being in pain', in V. Kestenbaum (ed.), *The Humanity of the Ill*, The University of Tennessee Press, Knoxville, Tennessee, pp. 101–124.

Schutz, A.: 1962a, 'Choosing among projects of action', in M. Natanson (ed.), *The Problem of Social Reality*, Vol. 1, *Alfred Schutz: Collected Papers*, Martinus Nijhoff, The Hague, pp. 67–96.

Schutz, A.: 1962b, 'Common sense and scientific interpretation of human action', in M. Natanson (ed.), *The Problem of Social Reality*, Vol. 1, *Alfred Schutz: Collected Papers*, Martinus Nijhoff, The Hague, pp. 3–47.

Schutz, A.: 1962c, 'Language, language disturbances and the texture of conscious-ness', in M. Natanson (ed.), *The Problem of Social Reality*, Vol. 1, *Alfred Schutz: Collected Papers*, Martinus Nijhoff, The Hague, pp. 260–286.

Schutz, A.: 1976a, 'Making music together', in A. Brodersen (ed.), *Studies in Social Theory*, Vol. 2, *Alfred Schutz: Collected Papers*, Martinus Nijhoff, The Hague, pp. 159–178.

Schutz, A.: 1962d, 'On multiple realities', in M. Natanson (ed.), *The Problem of Social Reality*, Vol. 1, *Alfred Schutz: Collected Papers*, Martinus Nijhoff, The Hague, pp. 207–259.

Schutz, A.: 1970, *Reflections on the Problem of Relevance*, R.M. Zaner (ed.), Yale University Press, New Haven, Connecticut.

Schutz, A.: 1962e, 'Some leading concepts in phenomenology', in M. Natanson (ed.), *The Problem of Social Reality*, Vol. 1, *Alfred Schutz: Collected Papers*, Martinus Nijhoff, The Hague, pp. 99–117.

Schutz, A.: 1962f, 'Symbol, reality and society', in M. Natanson (ed.), *The Problem of Social Reality*, Vol. 1, *Alfred Schutz: Collected Papers*, Martinus Nijhoff, The Hague, pp. 287–356.

Schutz, A.: 1976b, 'The homecomer', in A. Brodersen (ed.), *Studies in Social Theory*, Vol. 2, *Alfred Schutz: Collected Papers*, Martinus Nijhoff, The Hague, pp. 106–119.

Schutz, A.: 1975, 'The problem of transcendental intersubjectivity in Husserl', in I. Schutz (ed.), *Studies in Phenomenological Philosophy*, Vol. 3, *Alfred Schutz: Collected Papers*, Martinus Nijhoff, The Hague, pp. 51–91.

Schutz, A. and Luckmann, T.: 1973, *The Structures of the Life-World*, R.M. Zaner and H.T. Englehardt, Jr. (trans.), Northwestern University Press, Evanston, Illinois.

Schwartz, M.A. and Wiggins, O.: 1985, 'Science, humanism, and the nature of medical practice: A phenomenological view', *Perspectives in Biology and Medicine* 28, 331–361.

Schwartz, M.A. and Wiggins, O.: 1988, 'Scientific and humanistic medicine: A theory of clinical methods', in K.L. White (ed.), *The Task of Medicine, Dialogue at Wickenburg*, The Henry J. Kaiser Family Foundation, Menlo Park, California, pp. 137–171.

Schwartz, M.A. and Wiggins, O.: 1987, 'Typifications: The first step for clinical diagnosis in psychiatry', *Journal of Nervous and Mental Disease* 175, 65–77.

Silberman, C.: 1991, 'From the patient's bed', *Health Management Quarterly* **13**, 12–15.

Silberman, C.: Forthcoming, *Crisis in American Medicine*. Pantheon Books, New York.

Siller, J., *et al.*: 1976, *Structure of Attitudes Toward the Physically Disabled*, New York University School of Education, New York.

Simmel, M.: 1962, 'Phantom experiences following amputation in childhood', *Journal of Neurosurgery and Psychiatry* **25**, 69–72.

Sokolowski, R.: 1974, *Husserlian Meditations: How Words Present Things*, Northwestern University Press, Evanston, Illinois.

Sontag, S.: 1988, 'AIDS and its metaphors', *The New York Review of Books* **35**, 89–99.

Sontag, S.: 1978, *Illness as Metaphor*, Farrar, Straus and Giroux, New York.

Stein, E.: 1970, *On the Problem of Empathy*, W. Stein (trans.), Martinus Nijhoff, The Hague.

Stetten, D., Jr.: 1981, 'Coping with blindness', *The New England Journal of Medicine* **305**, 458–460.

Stoddard, S.: 1978, *The Hospice Movement: A Better Way of Caring for the Dying*, Stein and Day Publishers, New York.

Straus, E.: 1966, *Phenomenological Psychology, Selected Papers*, E. Eng (trans.), Basic Books, New York.

Sydenham, T.: 1981, 'Preface to the Third Edition: Observationes Medicae', in A.L. Caplan *et al.* (eds.), *Concepts of Health and Disease: Interdisciplinary Perspectives*, R.G. Latham (trans.), Addison-Wesley Publishing, Reading, Massachusetts, pp. 145–155.

Temkin, O.: 1981, 'The scientific approach to disease: Specific entity and individual sickness', in A.L. Caplan *et al.* (eds.), *Concepts of Health and Disease: Interdisciplinary Perspectives*, Addison-Wesley Publishing, Reading, Massachusetts, pp. 247–263.

Thorn, G.W. *et al.*: 1977, *Harrison's Principles of Internal Medicine*, 8th Edition, McGraw Hill, New York.

Tolstoy, L.: 1978, 'The Death of Ivan Ilych', in L. Perrine (ed.), *Story and Structure*, 5th Edition, Harcourt Brace Jovanovich, New York, pp. 502–544.

Toombs, S.K.: 1988, 'Illness and the paradigm of lived body', *Theoretical Medicine* **9**, 201–226.

Toombs, S.K.: 1992, 'The body in multiple sclerosis:A patient's perspective', in D. Leder (ed.), *The Body in Medical Thought and Practice*, Kluwer Academic Publishers, Dordrecht, The Netherlands.

Toombs, S.K.: 1987, 'The meaning of illness: A phenomenological approach to the patient-physician relationship', *The Journal of Medicine and Philosophy* **12**, 219–240.

Toombs, S.K.: 1990, 'The temporality of illness: Four levels of experience', *Theoretical Medicine* **11**, 227–241.

Toulmin, S.: 1976, 'On the nature of the physician's understanding', *The Journal of Medicine and Philosophy* **1**, 32–50.

Trenn, T.: 1981, 'Ludwik Fleck's 'On the Question of the Foundations of Medical

Knowledge'', *The Journal of Medicine and Philosophy* **6**, 237–256.

Updike, J.: 1976, 'From the journal of a leper', *The New Yorker*, 28–33.

Van den Berg, J.H.: 1972, *A Different Existence, Principles of Phenomenological Psychopathology*, Duquesne University Press, Pittsburgh, Pennsylvania.

Van den Berg, J.H.: 1955, *The Phenomenological Approach to Psychiatry*, Charles C. Thomas, Springfield, Illinois.

Virchow, R.: 1981, 'Three selections from Rudolf Virchow', in A.L. Caplan *et al.* (eds.), *Concepts of Health and Disease: Interdisciplinary Perspectives*, S.G.M. Engelhardt (trans.), Addison-Wesley Publishing, Reading, Massachusetts, pp. 187–195.

Watson, J. *et al.*: 1981, 'The diagnosis of abdominal pain in ambulatory male patients', *Medical Decision Making* **1**, 215–224.

Webster, B.: 1989, *All of a Piece: A Life With Multiple Sclerosis*, The Johns Hopkins University Press, Baltimore, Maryland.

Zaner, R.M.: 1985, 'A philosopher reflects: A play against night's advance', in P.D. Rabin *et al.* (eds.), *To Provide Safe Passage: The Humanistic Aspects of Medicine*, Philosophical Library, New York, pp. 222–246.

Zaner, R.M.: 1982, 'Chance and morality: The dialysis phenomenon', in V. Kestenbaum (ed.), *The Humanity of the Ill*, The University of Tennessee Press, Knoxville, Tennessee, pp. 39–68.

Zaner, R.M.: 1988, *Ethics and the Clinical Encounter*, Prentice Hall, Englewood Cliffs, New Jersey.

Zaner, R.M.: 1973a, 'Examples and possibles: A criticism of Husserl's theory of free-phantasy variation', *Research in Phenomenology* **3**, 29–43.

Zaner, R.M.: 1973b, 'The art of free phantasy in rigorous phenomenological science', in F. Kersten *et al.* (eds.), *Phenomenology, Continuation and Criticism: Essays in Memory of Dorian Cairns*, Martinus Nijhoff, The Hague, pp. 192–219.

Zaner, R.M.: 1981, *The Context of Self: A Phenomenological Inquiry Using Medicine as a Clue*, Ohio University Press, Ohio.

Zaner, R.M.: 1964, *The Problem of Embodiment: Some Contributions to a Phenomenology of the Body*, Martinus Nijhoff, The Hague.

Zaner, R.M.: 1970, *The Way of Phenomenology: Criticism as a Philosophical Discipline*, Western Publishing, New York.

Zaret, B.L.: 1987, 'Trauma', in H. Mandell *et al.* (eds.), *When Doctors Get Sick*, Plenum Publishing, New York, pp. 405–411.

# INDEX

159

# Philosophy and Medicine

1. H. Tristram Engelhardt, Jr. and S.F. Spicker (eds.): *Evaluation and Explanation in the Biomedical Sciences.* 1975                                                ISBN 90-277-0553-4
2. S.F. Spicker and H. Tristram Engelhardt, Jr. (eds.): *Philosophical Dimensions of the Neuro-Medical Sciences.* 1976                                         ISBN 90-277-0672-7
3. S.F. Spicker and H. Tristram Engelhardt, Jr. (eds.): *Philosophical Medical Ethics: Its Nature and Significance.* 1977                                       ISBN 90-277-0772-3
4. H. Tristram Engelhardt, Jr. and S.F. Spicker (eds.): *Mental Health: Philosophical Perspectives.* 1978                                                        ISBN 90-277-0828-2
5. B.A. Brody and H. Tristram Engelhardt, Jr. (eds.): *Mental Illness.* Law and Public Policy. 1980                                                               ISBN 90-277-1057-0
6. H. Tristram Engelhardt, Jr., S.F. Spicker and B. Towers (eds.): *Clinical Judgment: A Critical Appraisal.* 1979                                              ISBN 90-277-0952-1
7. S.F. Spicker (ed.): *Organism, Medicine, and Metaphysics.* Essays in Honor of Hans Jonas on His 75th Birthday. 1978                                         ISBN 90-277-0823-1
8. E.E. Shelp (ed.): *Justice and Health Care.* 1981
                                                                  ISBN 90-277-1207-7; Pb 90-277-1251-4
9. S.F. Spicker, J.M. Healey, Jr. and H. Tristram Engelhardt, Jr. (eds.): *The Law-Medicine Relation: A Philosophical Exploration.* 1981      ISBN 90-277-1217-4
10. W.B. Bondeson, H. Tristram Engelhardt, Jr., S.F. Spicker and J.M. White, Jr. (eds.): *New Knowledge in the Biomedical Sciences.* Some Moral Implications of Its Acquisition, Possession, and Use. 1982                                 ISBN 90-277-1319-7
11. E.E. Shelp (ed.): *Beneficence and Health Care.* 1982         ISBN 90-277-1377-4
12. G.J. Agich (ed.): *Responsibility in Health Care.* 1982       ISBN 90-277-1417-7
13. W.B. Bondeson, H. Tristram Engelhardt, Jr., S.F. Spicker and D.H. Winship: *Abortion and the Status of the Fetus.* 2nd printing, 1984    ISBN 90-277-1493-2
14. E.E. Shelp (ed.): *The Clinical Encounter.* The Moral Fabric of the Patient-Physician Relationship. 1983                                                 ISBN 90-277-1593-9
15. L. Kopelman and J.C. Moskop (eds.): *Ethics and Mental Retardation.* 1984
                                                                             ISBN 90-277-1630-7
16. L. Nordenfelt and B.I.B. Lindahl (eds.): *Health, Disease, and Causal Explanations in Medicine.* 1984                                                ISBN 90-277-1660-9
17. E.E. Shelp (ed.): *Virtue and Medicine.* Explorations in the Character of Medicine. 1985                                                              ISBN 90-277-1808-3
18. P. Carrick: *Medical Ethics in Antiquity.* Philosophical Perspectives on Abortion and Euthanasia. 1985                                ISBN 90-277-1825-3; Pb 90-277-1915-2
19. J.C. Moskop and L. Kopelman (eds.): *Ethics and Critical Care Medicine.* 1985
                                                                             ISBN 90-277-1820-2
20. E.E. Shelp (ed.): *Theology and Bioethics.* Exploring the Foundations and Frontiers. 1985                                                             ISBN 90-277-1857-1
21. G.J. Agich and C.E. Begley (eds.): *The Price of Health.* 1986
                                                                             ISBN 90-277-2285-4

# Philosophy and Medicine

KLUWER ACADEMIC PUBLISHERS – DORDRECHT / BOSTON / LONDON